The Inner Life of Animals

foreword by JEFFREY MOUSSAIEFF MASSON

PETER WOHLLEBEN

translation by JANE BILLINGHURST

The Inner Life of
ANIMALS

LOVE, GRIEF, AND COMPASSION
Surprising Observations of a Hidden World

DAVID SUZUKI INSTITUTE

GREYSTONE BOOKS

Vancouver/Berkeley

FRONT COVER ILLUSTRATION: Bärli, Snow, and Flocke, Peter Wohlleben's goats

Greystone Books Ltd.
www.greystonebooks.com

David Suzuki Institute
www.davidsuzukiinstitute.org

Cataloguing data available from Library and Archives Canada
ISBN 978-1-77164-301-6 (cloth)
ISBN 978-1-77164-302-3 (epub)

Copyediting by Lynne Melcombe
Jacket and interior design by Nayeli Jimenez
Jacket illustration by Briana Garelli
Printed and bound in Canada on ancient-forest-friendly paper by Friesens

We gratefully acknowledge the financial support of the Canada Council for the Arts, the British Columbia Arts Council, the Province of British Columbia through the Book Publishing Tax Credit, and the Government of Canada through the Canada Book Fund for our publishing activities.

Canadä

CONTENTS

FOREWORD

P ETER WOHLLEBEN'S PREVIOUS book, *The Hidden Life of Trees: What They Feel, How They Communicate— Discoveries from a Secret World,* became a sensation in Germany when it was first published in 2015, and it went on to equal success in many other countries, as well. With good reason. The book is a passionate, intelligent, lyrical account of the emotions (yes, you read that right) trees experience on their own and for others, and sometimes even for others of a different plant species (something we call altruism). Trees take care of other trees that are weak or sick, they cry out when they are thirsty, and they possess other astonishing, long-overlooked abilities. *The Hidden Life of Trees* was an update to the vastly popular but scientifically dubious book by the American World War II spy Peter Tompkins, published in 1973: *The Secret Life of Plants.* With Wohlleben, we were on the right track.

And now comes the same treatment given to the inner life of animals. The only difference is that this subject has, especially in the last ten years or so, taken off as a field of serious research and popular books, especially by Jonathan Balcombe, the oceanologist Carl Safina and many others. The trend was started, I believe, way back in 1982 by the great Harvard biologist Donald Griffin, who wrote the book *The Question of Animal Awareness.* These books have proved immensely popular precisely because the general population knows that what is being described is true from their own experiences. Who, living with a dog, or cat, or parrot, could ever doubt that these animals feel as much as we do? Or even more than we do, I would now venture after reading what Wohlleben has to say.

I believe, too, that this movement has led to a transformation in the dietary practices of many millions of people around the world. If we can safely eat alternative food, why cause unnecessary suffering, is the thinking, a sentiment with which I happen to agree completely. I am not entirely certain where Wohlleben comes out in this increasingly urgent debate, but he makes absolutely no secret of his profound loathing of sport hunting by providing a sense of what the animals actually feel. His writing is a *tour de force.*

Wohlleben's love of animals is apparent on every page of this absorbing book, and he has a special gift for allowing the rest of us to feel his enthusiasm and share in his knowledge and personal observations in a very simple way, all the while giving us good ammunition to demonstrate to skeptics how obvious his thesis is, when you stop and think about it. How far does he take it? Well, considering this is the man who has convinced the world that trees love their saplings, one has to be prepared for pushing the boundaries. Wohlleben does not disappoint. Ticks? You want to deny them any feelings? Well, then, be prepared for a counterargument from our author. I won't give it away, but everything he writes makes sense. That is one of his great strengths and what made his first book so adored: he says something completely out of left field, and within a few paragraphs he convinces you he is right and traditional science has been wrong for years.

Wohlleben convinced me over and over. Want to know the difference between the way hedgehogs and squirrels dream? Wohlleben will tell you. You will learn a lot about bees, because the author keeps them and he insists that "they plan for the future, reflect on things they have not yet seen, and are aware of their bodies in relation to these things" and so "they are self-aware." By the way, did you know that bees must visit 125,000 flowers to get enough nectar to produce a single ounce of honey? Gems like this are scattered throughout the book. Butterflies, too, come in for some interesting observations. (One of the great strengths of the book is how widely the author casts his net. Tardigrades, also known as water bears, anyone?) It is hard to say what animal he is most fond of, but I was particularly impressed with his many accounts of pigs and their sensitive nature. And, of course, their intelligence. When a law protected them from hunting in the canton of Geneva but not in neighboring France, as soon as the first shots of the season were fired, the wild boar swam across the Rhône to the safety of Switzerland. (I can't help wondering what kind of world we would live in if we never hunted.) Even when the territory is familiar, it is good to be reminded, for instance, that ravens and wolves form what appear to be friendships.

Now I know that a number of scientists will be up in arms about the way Wohlleben "proves" a thesis. Take this as an example: "A particularly moving example that animals, too, are capable of empathy across species lines comes from the Budapest Zoo. Aleksander Medveš was visiting the zoo and filming the brown bear in its enclosure when suddenly, a crow fell into the moat. The bird began to weaken as it thrashed about and was in danger of drowning, when the bear intervened. It carefully took one of the bird's feathers in its mouth and pulled the bird back to land. The bird lay there as though petrified, before it pulled itself together. The bear took no more notice of this fresh morsel of meat, which was definitely a potential prey item. Instead, it turned its attention once more to its meal of vegetables." I find this believable and I would love to have seen it, so I am grateful Wohlleben has made that possible. But I do see that a scientist will scream "anecdotal evidence." Wohlleben obviously doesn't care. Should we? Upon reflection, I think not. He is writing not as a scientist but as an observant animal lover who has put a great deal of thought into this charming, accessible book with its generous embrace of topics on which he sheds both light and warmth. Reading the book is pure delight.

JEFFREY MOUSSAIEFF MASSON

INTRODUCTION

ROOSTERS THAT DECEIVE their hens? Mother deer that grieve? Horses that feel shame? Up until just a few years ago, such ideas would have sounded absurd, mere wishful thinking on the part of animal lovers who wanted to feel closer to their charges. I've been around animals all my life and I was one of those dreamers. Whether it was the chick in my parents' garden that picked me out as its mom, the goat at our forest lodge that brightened our days with her contented bleating, or the animals I met on my daily rounds of the woodland I manage, I often wondered what was going on inside their heads. Is it really true, as scientists have long maintained, that people are the only animals capable of enjoying a full range of emotions? Has creation really engineered a unique biological path for us? Are we the only ones guaranteed a life of self-awareness and satisfaction?

If that were the case, this book would be over right now. If human beings were the result of some special biological design, we wouldn't be able to compare ourselves to other animals. It would make no sense to talk about empathy with them, because we would not be able to even begin to imagine how they felt. Luckily, Nature opted for the economy plan. Evolution "only" modifies and builds on whatever is already available, much like a computer system. And so, just as code from earlier operating systems is integrated into the latest Windows, the genetic programming of our ancient ancestors still works in us—and in all the other species whose family trees branched off from our lineage in the past few million years. And so, as I see it, there is only one kind of grief, pain, or love. It might sound presumptuous to say that a pig feels things just as we do, but there is a vanishingly small chance that an injury hurts a pig less than it hurts us. "Aha," the scientists might interject at this point, "but we have no proof." That's true, but there never will be any proof. I can't even prove that you feel the same way as I do. No one can look inside another person and prove that, say, a prick of a pin triggers the same sensation in each one of the seven billion people on this planet. But we are able to express our feelings in words, and this ability to share increases the probability that people operate on roughly the same level when it comes to feelings.

So when our dog Maxi polished off a bowl of dumplings in the kitchen and then looked up at us with an innocent expression on her face, she was not behaving like a biological eating machine; she was behaving like the shrewd and endearing little rascal she was. The more often and the more closely I paid attention, the more I noticed our pets and their wild woodland relatives displaying what are supposed to be exclusively human emotions. And I am not alone in this. More and more researchers are realizing that humans and many animals share things in common. True love among ravens? No question. Squirrels who know the names of their close relatives? That's been documented for a long time. Wherever you look, animals are out there, loving each other, feeling each other's pain, and enjoying each other's company.

Currently, there's a great deal of scientific research on the inner lives of animals, although it's usually so narrowly focused and written in such dry, academic language that it hardly makes for gripping reading and, more importantly, rarely leads to a better understanding of the subject. And that is why I would like to act as your interpreter and translate fascinating scientific research into everyday language for you, assemble the individual pieces of the puzzle so you can see the big picture, and sprinkle in a few observations of my own to bring it all to life. I hope this will help you see the animal world around you and the species described in this book, not as mindless automatons driven by an inflexible genetic code, but as stalwart souls and lovable rascals. And that is just what they are, as you will discover for yourself when you take a walk in my neighborhood with my goats, horses, and rabbits, or in the parks and woods where you live. Come on. I'll show you what I mean.

1

SELFLESS MOTHER LOVE

IT WAS A hot summer day at my forester's lodge deep in the woods near Hümmel in the Eifel, a mountain range in Germany. The year was 1996. To cool off, my wife and I had set out a wading pool under a shady tree in the garden. I was sitting in the water with my two children, and we were enjoying juicy slices of watermelon when, all of a sudden, I became aware of a movement out of the corner of my eye. A rusty-brown something was scampering toward us, freezing for an instant every now and then as it advanced. "A squirrel!" the children cried in delight. My joy, however, soon turned to deep concern as the squirrel took a few more steps and then keeled over onto its side. It was clearly ill, and after it had taken a few more steps (in our direction!), I noticed a large growth on its neck. It looked as though what we were dealing with here was an animal that was definitely suffering from

5

something and might even be highly infectious. Slowly but surely, it was approaching the pool. I was on the point of gathering up the children and beating a hasty retreat when the menacing advance resolved itself into a touching scene. The lump turned out to be a baby wrapped around its mother's neck like a furry ruff. The baby's stranglehold, along with the shimmering heat, meant the squirrel mother could only suck in enough air to take a few steps before falling over sideways, exhausted and gasping for breath.

A squirrel mother cares for her children with selfless devotion. When danger threatens, she carries them to safety in the manner I have just described. She can end up totally spent, because she may have as many as six tiny tots to tote one after the other, each one clasped around her neck. Despite her devotion, the chances of her little ones surviving are low, and about 80 percent die before they are a year old. Although the rusty rascals can avoid most enemies during the day, death stalks them at night while they are sleeping. When darkness falls, predatory pine martens creep through the branches to interrupt the squirrels' dreams. When the sun shines, the danger comes from agile hawks threading their way through the trees on the lookout for a tasty meal. When a hawk spots a squirrel, a spiral of fear begins. And I mean that literally, for the squirrel tries to escape the hawk by disappearing to the other side of the tree. The hawk banks steeply to follow its prey. In a flash, the squirrel disappears to the other side of the tree again. The hawk follows. Moving at breakneck speed, both animals spiral around the trunk. The nimbler one wins—usually the little mammal.

Winter, however, is more devastating than any predator. To make sure they go into the cold season well prepared,
squirrels build drays. They anchor these spherical nests
between branches high up in tree crowns and fashion two
separate exits with their paws so they can escape any uninvited guests. The nest is made mostly of small twigs, and
the interior is cushioned with soft moss that helps conserve
heat and provides a comfortable place to sleep. Comfortable?
Yes, animals value comfort, too, and squirrels don't like twigs
poking into their backs while they're trying to sleep any more
than we do. A soft moss mattress guarantees a restful night.

From my office window, I regularly see squirrels pulling
this soft green material from our lawn and carrying it high
up into the branches. And I see something else, too. As soon
as the acorns and beechnuts tumble to the ground in fall,
squirrels gather these nutrient-rich packages, carry them
a few yards, and bury them. They hide food caches to make
sure they have food over the winter. Instead of going into true
hibernation, they spend most of their winter days dozing. In
this state of winter lethargy, they use less energy than usual,
but they do not shut down completely like, say, hedgehogs.
Every once in a while, a squirrel wakes up and gets hungry.
Then it slips down the tree and looks for one of its numerous
caches. And it looks, and looks, and looks. At first, it's funny
watching the little animal trying to remember where it's hidden its food. It burrows a bit here, digs a little there, sitting
upright every once in a while as though taking a break to
think. But that doesn't help. The landscape has changed considerably since the fall. The trees and shrubs have lost their

leaves, the grass has dried up, and worse, everything might now be covered in cottony white snow. As the frantic squirrel continues its search, my heart goes out to it. Nature is ruthlessly sorting out who will live and who will die. Most of the forgetful squirrels—primarily this year's young—will not live to see spring because they will starve to death. Then I find small clumps of beech trees sprouting in the ancient beech preserves. Baby beeches look like emerald-colored butterflies fluttering at the ends of slender stalks and they usually grow alone. They gather in clumps only in places where a squirrel has failed to retrieve the nuts it stashed, often because it simply forgot where they were, with the fatal consequences I have just described.

I find the red squirrel to be a prime example of how we sort animals into categories. Their dark button eyes are adorable, their soft fur is a beautiful reddish color (there are also some that are brownish black), and they pose no threat to humans. In spring, young trees sprout from their forgotten food caches, so you could say they help establish new woodlands. In short, we are kindly disposed toward them. We avoid thinking about their favorite food: baby birds. From my office window at the lodge, I am also privy to their predatory raids. When a squirrel scales a tree in spring, consternation reigns in the small colony of fieldfares that raise their young in the old pines along the driveway. The little birds, which are related to thrushes, flutter around the trees, chittering and chattering, trying to drive off the intruder. Squirrels are the birds' deadly enemies, because the little mammals calmly help themselves to one downy chick after another. Even nesting cavities offer the baby birds only limited protection. Armed with long, sharp claws at the end of slender paws, squirrels can fish even supposedly well-protected nestlings out of the tree hollows where they are hiding.

So are squirrels bad or are they good? Neither. A quirk of Nature ensures that they arouse our protective instincts, and so we experience positive emotions when we see them. This has nothing to do with them being good or useful. And on the flip side of the coin, their habit of killing the songbirds we also love doesn't mean they are bad either. The squirrels are hungry and must feed their young, which depend on nourishing milk from their mother. We would be thrilled if squirrels met their need for protein by gorging themselves on the caterpillars of the cabbage white butterfly. If they did this, our emotional balance sheet would come out 100 percent in the squirrels' favor, because these pests are a nuisance in our vegetable gardens. But caterpillars are also young animals, and in this case they grow up to be butterflies. And just because the caterpillars happen to like the plants we have earmarked for our dinner doesn't mean that killing butterfly babies counts as a net benefit for the natural world. The squirrels, meanwhile, are not the slightest bit interested in what we think of them. They are too busy surviving and, while they are at it, making the most of life.

But back to maternal love in these little red rascals. Are they really capable of experiencing this emotion? A love so strong that a squirrel mother places a higher value on the lives of her offspring than she does on her own? Isn't it just a case of a spike in the hormones coursing through the squirrel's veins that triggers preprogrammed protective behavior? Science has a tendency to reduce biological processes to involuntary mechanics, so, before painting such a dispassionate picture of squirrels and other animals, let's take a look at maternal love in our own species. What happens in a human mother's body when she holds an infant in her arms? Is maternal love innate? Science would say yes and no. Maternal love itself is not innate, but the conditions necessary for developing this love are.

Shortly before a child is born, the hormone oxytocin flows through the mother's system, which helps her develop a strong bond with her child. In addition, large quantities of endorphins are released, which dull pain and reduce anxiety. This cocktail of hormones remains in the mother's bloodstream after the birth of her child, ensuring that the baby is welcomed into the world by a mother who is relaxed and in a positive mood. Nursing stimulates further production of oxytocin, and the mother-child bond intensifies. The same thing happens in many animals, including the goats my family and I keep at our forest lodge (goat mothers also produce oxytocin). A mother goat starts getting acquainted with her kids when she licks off the mucus that covers her babies after birth. The clean-up process intensifies their bond, and as the mother goat bleats softly to her children, her offspring reply in thin, reedy voices, and the vocal signatures are imprinted in both mother and kids.

Things do not go well if something goes awry at clean-up time. When a mother goat in our small herd is ready to give birth, we put her in a stall of her own so she can deliver her kids in peace. There is a small gap under the door of the stall, and once during a birth, a particularly small kid slipped out under it. By the time we noticed the mishap, precious time had passed, and the mucus covering the kid had already dried. The result? Despite our best efforts, the mother goat refused to accept her baby. The time to trigger mother love had passed.

Something similar can happen with people. If a human mother in hospital is separated from her newborn baby for an extended period of time, the maternal bond becomes more difficult to establish. The situation is not as dramatic as with goats, because people are not totally dependent on hormones and can learn how to love. If people were like goats, adoptions would never work out, because adoptive mothers often meet their children years after their birth. Adoption, therefore, is the best opportunity we have for investigating whether maternal love is more than just an instinctive reflex and something that can be learned. But before we tackle this question, I would like to shine some light on instincts and how they work.

2

INSTINCT—
A SECOND-RATE EMOTION?

OFTEN HEAR THAT there's no point comparing animal emotions to human emotions, because animals act and feel instinctively, whereas humans act consciously. Before we turn to the question of whether instinctive behavior is second-rate, let's take a closer look at instincts. Science uses the term "instinctive behavior" to describe actions that are carried out unconsciously without being subjected to any thought processes. These actions can be genetically hard wired or they can be learned. What is common to all of them is that they happen very quickly because they bypass cognitive processes in the brain. Often these actions are the result of hormones released at certain times (in moments of anger, for example), which then trigger physical responses. So are animals nothing more than biological automatons on autopilot?

Before rushing to judgment, let's consider our own species. We are not free of instinctive behavior ourselves. Quite the opposite, in fact. Think about a hot element on an electric stove. If you were to absent-mindedly put your hand on one, you'd take it away again in a flash. There's no preceding conscious reflection, no internal monologue along the lines of: "That's strange. It smells like someone's barbecuing something and my hand suddenly really hurts. I had better remove it." You just react automatically without making a conscious decision to remove your hand. So people behave instinctively, too. The question is simply the extent to which instincts determine what we do every day.

To shed some light on the matter, let's turn to recent studies of the brain. The Max Planck Institute in Leipzig published the results of an astonishing study carried out in 2008. With the help of magnetic resonance imaging (MRI), which translates brain activity into digital images, test subjects were observed making decisions (whether to push the computer button with their right hand or with their left). The activity in their brains clearly showed what their choices were going to be up to seven seconds before the test subjects themselves were aware of them. This means that the behavior had already been initiated while the volunteers were still considering what to do. And so it follows that it was the unconscious part of the brain that triggered the action. It seems that what the conscious part of the brain did was to come up with an explanation for the action a few seconds later.[1]

Research into these kinds of processes is still very new, and so it's impossible to say what percentage and what kinds of decisions work this way, or whether we're capable of rejecting processes set in motion unconsciously. But still, it's amazing to think that so-called free will is often playing catch-up. All the conscious part of the brain is doing in this case is coming up with a face-saving explanation for our fragile ego, which, thanks to this reassurance, feels it's completely in control at all times. In many cases, however, the other side— our unconscious—is in charge of operations.

In the end, it doesn't really matter how much our intellect is consciously in control, because the fact that a surprising number of our reactions are probably instinctive shows only that experiences of fear and grief, joy and happiness are not at all diminished by being triggered instinctively instead of being actively instigated. Their origin doesn't reduce their intensity in any way. The point is that emotions are the language of the unconscious, and in day-to-day life, they prevent us from sinking beneath an overwhelming flood of information. The pain in your hand when you put it on a hot element allows you to react immediately. Feeling happy reinforces positive behaviors. Fear saves you from embarking on a course of action that could be dangerous. Only the relatively few problems that actually can and should be solved by thinking them through make it to the conscious level of our brain, where they can be analyzed at leisure.

Basically, then, emotions are linked to the unconscious part of the brain, not the conscious part. If animals lacked consciousness, all that would mean is that they would be unable to have thoughts. But every species of animal experiences unconscious brain activity, and because this activity directs how the animal interacts with the world, every animal must also have emotions. Therefore, instinctive maternal love cannot be second-rate, because no other kind of maternal love exists. The only difference between animals and people is that we can consciously activate maternal love (and other emotions)—for example, in the case of adoption, where there can be no question of an instinctive bond created between mother and child at birth because first contact often happens much later. Despite this, instinctive maternal love develops over time, and when it does, the accompanying hormone cocktail flows through the mother's bloodstream.

Aha! Have we finally successfully isolated a human emotional domain that animals cannot enter? Let's take another look at our red squirrel. Canadian researchers have been watching its relatives in the Yukon for more than twenty years. About seven thousand animals took part in the study, and, although red squirrels are solitary animals, five adoptions were observed. Admittedly, each case involved squirrel babies of a close family member being raised by another female. Only nieces, nephews, or grandchildren were adopted, which shows that squirrel altruism has its limits. From a purely evolutionary standpoint, there are advantages to this arrangement, because it means very closely related genetic material is preserved and handed down. Although it has to be said that five cases in twenty years is not exactly overwhelming proof of an adoption-friendly attitude in squirrels.[2] So let's take a look at some other species.

What about dogs? In 2012, a French bulldog called Baby hit the headlines. Baby lived in an animal sanctuary in

Brandenburg, Germany. One day, six baby wild boar were brought in. The sow had likely been shot by hunters, and the tiny striped piglets wouldn't have stood a chance on their own. At the sanctuary, the animals got full-fat milk—and full-on love. The milk came from the caregivers' bottles, but the love and warmth came from Baby. The bulldog adopted the whole crew right away and allowed the piglets to sleep snuggled up to her. She also kept a watchful eye on the little tykes during the day.[3] But could that be called a true adoption? After all, Baby didn't nurse the piglets. But nursing is not a necessary component of human adoptions either, and yet there are reports of dogs—such as the Cuban dog Yeti—who even did that. Yeti had just given birth to a litter of puppies, which meant she had a lot of milk. When a few pigs on the farm also had babies, Yeti lost no time adopting fourteen piglets, even though their own mothers were still around. The little piglets followed their new mom around the farmyard and, of most importance here, Yeti nursed them.[4] Was that an example of conscious adoption? Or did Yeti just have maternal instincts to spare? We could ask these same questions of human adoptions, where people with strong desires look for and find an outlet for them. You could even liken the keeping of dogs and other pets to interspecies adoption; after all, some four-legged friends are accepted into human society almost as though they were members of the family.

There are other cases, however, where superabundant hormones or surplus milk can be ruled out as the driving forces behind adoption. The crow Moses is a touching example. When birds lose their brood, Nature gives them another opportunity to work off their pent-up impulses. They can simply start anew and lay another clutch of eggs. There's no way a single bird like Moses can exercise its maternal instincts, yet Moses attempted to do just this. The target of Moses's attention was a potential enemy—a housecat—albeit an extremely small and relatively helpless one, because the kitten had obviously lost its mother and had not had anything to eat in a long time. The little stray popped up in Ann and Wally Collito's yard. The couple lived in a cottage in North Attleboro, Massachusetts, and they watched in amazement at what happened next. A crow attached itself to the little orphan and was clearly looking after it, feeding it with earthworms and beetles. Of course, the Collitos didn't just stand by and watch; they fed the kitten, as well. The friendship between the crow and the cat continued after the cat grew up, and it lasted until the crow disappeared five years later.[5]

But let's get back to instincts. In my opinion, it makes no difference whether a mother's love is triggered by unconscious commands or comes after conscious deliberation. At the end of the day, it feels just the same. What is clear is that people are capable of both, although instinctive love triggered by hormones is more common. Even if animals are not capable of consciously developing maternal feelings (and the adoption of animals across species barriers should make us rethink that one), instinctive maternal love remains, and it is just as moving and just as compelling. The squirrel that crossed our lawn in a haze of heat with a baby wrapped around her neck was motivated by deep devotion. And, when I think back on that day, knowing that makes the experience all the more beautiful.

3

LOVING PEOPLE

C AN ANIMALS REALLY love us? We've already seen in the case of squirrels just how difficult it can be to verify this feeling between animals of the same species. But to now add love across the species divide—and all the way to us? You've got to wonder whether this is simply wishful thinking to make it easier for us to justify imprisoning our pets. First, let's take another look at the mother-child bond, because this particularly strong kind of love is something we actually can trigger in animals, as I experienced when I was a boy.

Even back then, my interests revolved around Nature and the environment, and I spent every spare moment outside in the woods or at lakes in abandoned quarry pits along the Rhine. I imitated the calls of frogs to get them to respond, kept a few spiders in glass jars so I could observe them, and raised mealworms in flour to watch them turn into black beetles. In the evening, I curled up with books about behavioral biology (don't worry, books by adventure writers such as Karl May and Jack London also had their place by my bed). In one of them, I read that you could get chicks to imprint on people. All you had to do was incubate an egg and talk to it just before it hatched so that the tiny creature inside became imprinted on a human instead of a hen. Apparently, the relationship lasted a lifetime. How exciting!

At the time, my father kept a few hens and a rooster in the garden, so I had access to fertilized eggs. I didn't have an incubator, so an old electric blanket had to do. There was one other problem. Chicken eggs need to be kept between 100 and 101 degrees Fahrenheit and turned often every day so they can cool down a bit. Armed with a scarf and a thermometer, I had to painstakingly simulate behavior that comes naturally to a hen. For twenty-one days, I measured the egg's temperature, draped varying layers of scarf over it, and carefully turned it. A few days before the estimated date of hatching, I began my monologues. And then it happened. Punctually on the twenty-first day, a small packet of fluff pecked its way to freedom. I immediately christened it Robin Hood.

The chick was incredibly adorable. Its yellow feathers were sprinkled with tiny black spots, and its black button eyes gazed straight at me. It never wanted to leave my side, and every time it lost sight of me, it began to cheep frantically. It didn't matter if I was on the toilet, in front of the television, or in bed, Robin was always there. The only time I left the little chick alone was when I went to school. Then I took my leave with a heavy heart, and I was greeted effusively when I returned. But this intimate bond began to stress me out. My brother took pity on me and cared for the chick part time so that I could do something without Robin every once in a while. Eventually, however, it became too much for him, as well. By now, Robin had developed into a young hen, and we gave her to a retired English teacher who was very fond of animals. Man and hen became fast friends, and for a long time you could see the two of them taking walks together in the neighboring village: the teacher on foot with Robin riding on his shoulder.

I think it's safe to say that Robin established a genuine relationship with her human caregivers, and many people can share similar stories about being a substitute parent for a young animal. The bottle-fed kids my wife hand-raises, for example, remain extremely attached to her for life. Here and in other cases, human caregivers play the role of adoptive mothers, and the stories are always heart-warming. However, these relationships are not voluntary, at least not as far as the animals are concerned, even if they do have their caregivers to thank for their survival. It would be more meaningful if an animal were to come and stay with us of its own accord. But has this ever happened?

To find this out, we must leave the warm embrace of maternal love and cast a wider net. What we're looking for is a scenario where an animal can grow up and decide for itself whether it will stay or leave. There's a good reason most dogs and cats come to us as babies, because that removes the element of choice for the little scamps. And that's absolutely a good thing. After a few days of getting used to their new circumstances—and possibly after a twinge of anxiety at being separated from their mother—young animals just a few weeks old quickly get attached to their caregivers, and exactly like my wife's bottle-fed kids, they remain particularly close to their people for as long as they live. Everyone feels good, but there's still that nagging question: are there any adult animals that enter into relationships with people of their own free will?

For house pets, the answer is a resounding yes. There are countless examples of stray cats and dogs that practically force themselves on caring humans. But in answering this question, I'd prefer to explore the world of wild animals, because wild animals have not had tameness bred into them and are therefore not predisposed to seeking a connection with people. And I'd like to reject one more scenario: using food to tame animals, because when wild animals are offered food the only thing they want to do is eat, and therefore they tolerate, and to a certain extent get habituated to, our presence. Our former neighbors found out what a nuisance this can be when they started feeding a squirrel. For weeks, they had been tempting the little rascal with nuts, and it had practically become a member of the family. But if the human food dispenser wasn't there in a timely manner every day, the squirrel would start scratching impatiently at the window. It demolished the frame in just a few weeks; squirrel claws are razor sharp.

Most friendships between wild animals and people are to be found in the ocean, with dolphins. Fungie, who lives in Dingle Bay in Ireland, is a particular star. He pops up often, accompanies tour boats, and shows off for visitors. He's become a real tourist magnet and features in official travel brochures. People who feel moved to do so can safely get into the water with him. The sizable dolphin swims alongside them, and they experience a special kind of joy in his presence. His tameness doesn't depend on food, which he refuses to accept. Fungie has been around for over thirty years now, and it's difficult to imagine life in Dingle without him. Most people find Fungie's story delightful—but not everyone. A reporter for the German newspaper *Die Welt* interviewed scientists and asked whether the dolphin might not simply be deranged. Perhaps, the reporter asked, the solitary animal hangs out with people only because he's shunned by others of his kind?[6]

Apart from the fact that people often form friendships with animals for similar reasons—for example, because they are lonely after the loss of a partner—I would like to investigate the question further with land-based animals closer to home. And that's not easy, because a common characteristic of wild animals is that they are exactly that—wild—and therefore they normally never seek contact with people. Moreover, people have hunted them for tens of thousands of years, so they have evolved to be wary of people; those that don't escape in time are in danger of losing their lives. And that is still the case for many animals, as you can see just by running your eye down the list of animals it's still legal to hunt. Whether they are large game such as deer or wild boar, or smaller four-footed targets such as foxes or hares, or even birds, from raptors to geese and ducks or snipe, every year thousands upon thousands meet their end in a hail of bullets. Thus, a certain mistrust of anything on two legs is completely understandable. And that is why we are so moved when such a creature overcomes its natural wariness and seeks contact with us.

What might motivate a wild animal to do that? Let's dismiss attracting them with food, because then we don't know whether it's just a case of hunger overriding fear. There is another driving force, however—one that is important for people as well—and that is curiosity. My wife, Miriam, and I had the good fortune to encounter at least one curious species: reindeer in Lapland. Okay. The reindeer are not completely wild, because the indigenous people, the Sami, own the animals and herd them with helicopters and all-terrain vehicles when they want to sort them for butchering or branding. Despite this, the reindeer have retained their wild character and are usually very wary around people.

Miriam and I were tent camping in the mountains in Sarek National Park, and because I am an early riser by nature, I was the first to creep out of my sleeping bag in the morning. I had been gazing for a while at the breathtaking sight of Nature untouched by human hands when I suddenly became aware of movement close by. A reindeer! Just the one? No, there were more coming down the slope, and I woke Miriam so she could watch the animals, as well. As we ate breakfast, more and more reindeer gathered round, until we were surrounded by the whole herd—about three hundred animals. The reindeer spent all day around our tent, and one young calf even dared to get within a few yards so it could lie down by the tent for a midday nap. We felt we were in paradise.

When a small group of hikers walked by, we realized how wary of people these animals really were. As soon as the hikers appeared, the herd retreated, only to return a while later to the area around our tent. It was clear that some of them were very interested in us. Eyes wide open and nostrils flared, they tried to figure us out. For us, it was the most amazing experience of the whole trip. We have no idea why the reindeer were so trusting around us. Perhaps our body language is calmer than usual for humans because of our daily interactions with animals, and that made us seem less threatening.

Anyone can have similar interactions in places where animals are not hunted. In national parks in Africa, for instance, or on the Galapagos Islands, or out on the tundra in the far north—places where species have not yet had bad experiences with people—animals allow visitors to get very close to them. And every once in a while, there are some individuals who are curious enough to want to check out the unusual guests wandering about in their territory. These are the encounters that make people particularly happy, because both parties come to them completely voluntarily.

It is difficult to prove that an animal truly loves a person of its own free will. Even my little chick, Robin Hood, had no real choice but to develop feelings for me. How about looking at it the other way around? Every owner of a pet, be it a cat or a dog or some other animal, knows that people are capable of loving animals. But what about the quality of this love? Some might argue that people simply project their emotions onto animals and see them reflected back. Their pets are substitutes for children they wish they'd had, partners they've lost, or friends who keep their distance. The subject is a mine-field that I would just as soon avoid; however, as we're talking about animal emotions, we should ask how our sentimental attachments affect our four-legged friends.

First off, they literally deform animals. In most places in the world, it's been a long time since cats and dogs were bred to be highly skilled helpers in hunting hares, deer, or mice. Instead, we've been breeding them to satisfy, in both char-acter and appearance, our desire to have something to cuddle and hug. The French bulldog is a good example. I used to think they were ugly, and that their squashed, wrinkled snouts put them at a disadvantage, because their snub noses interfere with their breathing so much that they snore. But then I got to know Crusty, a blue-gray male that we looked after every once in a while. Crusty won me over right away, and from that moment on, I no longer cared how he had been bred—he was just so adorable. Whereas other dogs have had enough after five minutes of being stroked, Crusty enjoyed this treatment for hours. If you stopped, he would nudge your hand beseech-ingly and look at you with his big puppy-dog eyes. His favorite activity was snoring contentedly while sleeping on his own-er's stomach.

Can breeding like that be a bad thing for the dog? There's no question that French bulldogs have been bred to be lap dogs—living cuddly toys, so to speak. I don't want to judge the legitimacy of this. The more important question I'd like to ask is what is this like for the dog? If a heightened need to be stroked has been bred into it, and if its appearance causes everyone (and I mean everyone!) to want to satisfy this need immediately, does the dog have a problem? It obviously feels just fine, and both it and the people it meets get what they wish for. It's just that what led to this need to be stroked—genetic manipulation through selective breeding—has a tiny trace of the unnatural about it. This is very different from cases where owners ignore their animals' needs, whether natural or caused by breeding, and when self-interested love blinds them so much that they end up treating their pets like people dressed in dog costumes. In such cases, overfeeding, insufficient exercise, and lack of exposure to the delights of the outdoors (such as walks in the snow) lead to severe health problems that torture the pampered animals to death.

4

ANYBODY HOME?

BEFORE WE DELVE deeper into the emotional and inner life of animals, we should ask once again whether the idea is just too far-fetched. After all, we need to have certain brain structures in order to process the emotions we experience, at least according to current thinking in science. The answer is pretty clear: in people, it is the limbic system that allows us to experience the full range of joy, grief, fear, or desire and, together with other areas of the brain, it facilitates the appropriate physical reactions.[7] These brain structures are very old in evolutionary terms and so we share them with many mammals. Goats, dogs, horses, cows, pigs—the list goes on and on. According to recent research, not only mammals but also birds and even fish, which biologists rank far lower on the evolutionary scale, belong on this list.

In the case of aquatic animals, it was pain research that led to the topic of emotions. The starting point was fishing and whether fish can feel the injuries caused by hooks. What might appear self-evident to you was thought to be unlikely for a long time. When you see photographs of fishing trawlers pulling aboard nets filled with living, slowly suffocating ocean inhabitants, when you see a trout thrashing around at the end of an angler's bent rod, you have to ask yourself how society tolerates such behavior in light of today's discussions about animal welfare. It's probably not a case of intentional ill will, but rather acceptance of the mostly unproven assumption that fish are witless creatures that swim around in rivers and oceans not feeling anything at all.

Victoria Braithwaite, a professor at Penn State who earned her doctorate from the University of Oxford, discovered something quite different. Years ago, she identified more than twenty pain receptors around fishes' mouths right where anglers' hooks usually catch hold.[8] Ouch! But all that proves is that a dull feeling of pain is within the realm of possibility. And so Braithwaite poked needles into the areas she had identified, which caused a reaction in the fishes' hindbrain, which is exactly where pain stimuli are processed in people. That should be enough to prove that fishing injuries cause fish to suffer.

But what about emotions? Let's consider fear. In people, fear arises in the almond-shaped amygdala. For a long time that was not proven, even though it was suspected. It wasn't until January 2011 that scientists from the University of Iowa published a paper about a woman identified as SM. SM was afraid of spiders and snakes—until the cells in her amygdala died after a rare illness. That was, of course, tragic for SM;

however, it offered researchers a unique opportunity to investigate what happens when this organ is lost. They took SM to a pet shop and confronted her with the animals she feared. The woman could now touch the animals, something she could never have done before, and she reported that she simply felt curious about them and no longer felt in the least bit frightened.[9] And so the seat of fear can now be precisely located in people. But what about fish?

Manuel Portavella García and his team at the University of Seville have indeed found comparable structures in the outer areas of the fish brain, a place where no one had looked before. (In people, the fear center lies deep inside the lower part of the brain.) First, the researchers trained goldfish to swim quickly away from one corner of their tank as soon as a green light came on. If they failed to do this, they got an electric shock. Then, the researchers disabled a part of the fishes' brain known as the telencephalon. It corresponds with our fear center, and switching it off had the same result as in people: from then on the goldfish were no longer afraid of the green light and they ignored it. From this, the researchers concluded that fish and terrestrial vertebrates have inherited the same brain structures from common ancestors that lived more than 400 million years ago.[10]

It follows that all vertebrates have had the hardware for emotions for a very long time, but does that mean that animals feel things the same way we do? Much points in that direction. Scientists have even found oxytocin in fish, the hormone that not only brings joy to mothers, but also strengthens the love between partners. Joy and love in fish? We're not going

THE INNER LIFE OF ANIMALS

to be able to prove that, at least not in the foreseeable future, but if there's any doubt, why do we assume they are guilty as charged? Scientists have spoken out against emotions in animals for so long that their view is mostly accepted, but wouldn't it be better to give animals the benefit of the doubt to be sure that they are not suffering unnecessarily?

In the preceding chapters, I have purposely described emotions as we experience them. This is the only way we might be able to understand what's going on inside animals' heads. But even if the structures in their brains differ from ours and these differences mean that they probably experience things differently, that certainly doesn't mean that emotions in animals are inherently impossible. It simply means that it is more difficult for us to imagine what their emotions might feel like. Take the fruit fly, for example, whose central nervous system is made up of 250,000 cells, making it one four-hundred-thousandth the size of ours. Can such minute creatures with such a limited capacity up top really feel anything? Can they even be said to possess consciousness—this being, of course, the pinnacle of achievement. Unfortunately, science is not yet advanced enough to be able to answer this question, partly because the concept of consciousness cannot be clearly defined.

The closest we can get to a definition is that consciousness involves thinking and reflecting on things that we have experienced or have read about. Right now, you're thinking about what you're reading, and so you possess consciousness. And at a very basic level, the conditions necessary for consciousness have been discovered in the tiny brains of fruit flies. The flies are constantly barraged by stimuli from the external world, just as we are. The smell of roses, car exhaust, sunlight, a breath of air—all are registered by a variety of unconnected nerve cells. So how does a fly filter out from this flood of sensations what is most important so that it can stay out of danger and not miss out on a particularly tasty morsel? Its brain processes the information and ensures that different areas coordinate their activities, strengthening certain stimuli. And so what is of interest stands out from the general noise of thousands of other impressions. The flies, therefore, can focus their attention on specific things—just as we can.

Fruit flies' eyes are made up of about six hundred individual facets. Because these tiny insects dart around so quickly, their eyes are bombarded by a huge number of images per second. This seems like an impossibly large amount of data to process, but the flies must do this if they are to survive. Anything that moves could belong to a voracious predator. Therefore, the fruit fly brain leaves all static images blurry and focuses exclusively on moving objects. You could say that the tiny tykes are stripping things down to the bare essentials, an ability that you surely would not have expected these little flies to have. By the way, we do something similar. Our brain doesn't allow all the images we see to make it through to our consciousness. It only lets the important ones through. Does that mean flies have consciousness? Researchers won't go that far; however, it is clear that flies can at least actively focus their attention on what matters most to them.[11]

Let's return to variations in brain structure between species. The basic organ is certainly present even in lower vertebrates, but for the quality of feelings we experience, more is needed. You read over and over again that intense emotions of which the subject is consciously aware are only possible with central nervous systems like ours. The stress lies on the words "consciously aware." The grooves and ridges in our brain occur in its outermost layer, the neocortex, which is the most recent part of our brain to evolve. This is the seat of self-awareness and consciousness, the place where thinking happens. And the human brain has more of these cells than any other species. The crowning achievement of creation sits there right under our skull. It follows that all other creatures must be less aware of the constellation of emotions we experience and cannot be as intelligent as we are, right? Consider comments made by Germany's first professor of fisheries and fishing, Robert Arlinghaus, co-author of a study on pain in fish for the German government. In an interview with the German magazine *Spiegel Online,* he stressed that fish cannot experience pain like we do from the injuries they suffer when they are caught because they do not have a neocortex and therefore they can have no conscious awareness of pain.[12] Apart from the fact that other scientists do not agree with him (see below), this sounds to me more like a rationalization of his hobby than a reasoned, objective scientific opinion.

Gourmets advance a similar argument every year at Christmas when it's customary to load the table with tasty crustaceans, and *Der Spiegel* (the print sister magazine to *Spiegel Online*) has reported on this, as well.[13] The poster child for the spectrum of shellfish is the lobster, which is served up on a platter as a status symbol after being boiled bright red. Boiled alive, that is. Whereas vertebrates are killed before they are cooked, it's perfectly acceptable to throw crustaceans into a bubbling pot with all their senses intact. It can take minutes until the heat makes its way completely inside the animal, destroying its sensitive nerve endings. Pain? How can that be? Crustaceans don't have a spinal cord, and that means they are incapable of feeling pain. Or, at least, that is what people say. Their nervous system is configured differently from ours, and it's even more difficult to prove pain in crustaceans than it is in species that have an internal skeleton like we do. Scientists arguing on behalf of the food industry insist that the animals' reactions are nothing more than reflexes.

Professor Robert Elwood at Queen's University in Belfast disagrees. "Denying that crabs feel pain because they don't have the same biology is like denying they can see because they don't have a visual cortex [the part of the brain responsible for sight in people]."[14] Apart from that, pain is a component in reflex actions, as you can easily test for yourself if there's an electric fence nearby. If you put your hand on it and get an electric shock, you have no choice but to pull back right away, whether you want to or not. It's pure reflex on your part, something you do without a moment's thought, but that doesn't make the electric shock any less painful.

Is there really only one way—the human way—to experience feelings intensely and perhaps consciously? Evolution is not the single-track process we sometimes think (or maybe hope) it is. And birds, some of which possess diminutive brains, are a prime example that there is more than one route to intelligence. Since the age of their ancestors, the dinosaurs, their development has followed a different path from ours. Even without a neocortex, they can perform mental feats of the highest order. In birds, a region called the dorsal ventricular ridge oversees similar tasks and functions as our cerebral cortex. In contrast to the human neocortex, which is built up of layers, the equivalent area in the bird brain is made up of small clumps, a fact that fed long-standing doubt that it could perform a similar function.[15] Today we know that ravens and other species that live in social groups can match, and in some cases even exceed, the mental prowess of primates. This is further proof of science's practice of arguing too cautiously when in doubt about feelings in animals, denying them many mental capacities until there is positive proof that they possess them. Instead, couldn't we simply (and just as accurately) say: "We don't know"?

Before I end this chapter, I would like to introduce you to one more creature in our woods, an organism that is mindless in the truest sense of the word. Sometimes you can find it on rotting wood, where it forms a small, bumpy, yellow mat. It's a fungus. Hold on a moment. Isn't this a book about animals? Well, in the case of this fungus, science is not exactly sure which category it belongs to. It's difficult enough with normal fungi, which form a third kingdom of living things in between animals and plants, because they cannot be clearly assigned to either category. Like animals, fungi subsist on organic substances from other living beings. In addition, their cell walls are made of chitin, like the exoskeletons of insects. And the slime mold that creates that yellow mat on dead wood can even move. At night, like gelatinous jellyfish, these organisms are capable of slithering out of the glass lab containers where they are temporarily confined. Today, science is moving them out of the realm of fungi and edging them a step closer to animals. Welcome to this book.

Researchers find some kinds of slime molds so interesting that they regularly observe them in the laboratory. *Physarum polycephalum*, with its somewhat awkward Latin name, is just such a customer, and it loves rolled oats. Basically, the creature is one giant cell with countless nuclei. What researchers are now doing is placing these slimy unicellular organisms in a maze with two exits and putting food at one of the exits as a reward. The slime mold spreads out into the maze and after a hundred hours or more, finds the exit with the oats—not bad, really. To do this, it clearly uses its own slime trail to recognize where it has already been. It then avoids those areas because they have not led to success. In nature, such behavior is of practical benefit, because the creature knows where it has already been in its search for food and, therefore, the places where there isn't any food left. It's quite a feat to be able to solve a maze when you don't have a brain, and researchers credit these moving mat-like creatures with having some kind of spatial memory.[16] Japanese researchers topped it all off by using a slime mold to reproduce a map of the most important transportation routes in Tokyo. To do this, they set a slime mold down on a damp surface at a point that represented the center of the city. Piles of food marked the principal neighborhoods as attractive places to visit. The slime mold set off, and when it had connected the neighborhoods using the optimal, shortest route, there was a big surprise. The image pretty much corresponded to the suburban train system in the metropolis.[17]

I particularly like the slime mold example, because it shows how little it takes to overturn our preconceptions about primitive Nature and stupid, emotionless animals. These alien creatures lack any of the basics laid out in the preceding chapters, and yet, if organisms with only a single cell have spatial memory and can perform complex tasks, how many undreamt-of skills and emotions might there be in animals with as few as 250,000 brain cells, like the fruit flies I've just introduced you to? Given how much more like us birds and mammals are in the physical structure of their bodies and brains, it would hardly come as a surprise if we were to discover that they are as sensitive to the world as we are.

5

PIG SMARTS

OMESTIC PIGS ARE descended from wild boar, which were prized by our ancestors as a source of meat. About ten thousand years ago, wild boar were tamed to ensure the delicious animals were available at short notice without us having to go out on dangerous hunts to get them, and they were then bred to better satisfy our requirements. Despite this interference, modern domestic pigs have retained wild boar's behavioral repertoire and, above all, their intelligence.

First, let's look at how wild boar behave. (Feral hogs or swine in the US are descended from pigs that escaped domesticity and share behaviors with European wild boar.)[18] For example, wild boar know exactly which other boar they are related to, even if the connection is a distant one. Researchers from Dresden University of Technology determined this indirectly when they were investigating the home ranges of family groups (known as sounders). As part of this research, 152 wild boar were caught in traps or stunned with tranquilizer guns, fitted with transmitters, and then set free again so the researchers could see where these nocturnal roamers hung out. The researchers discovered that there is normally not much overlap between neighboring sounder territories, and, on average, territories range from 1½ to 2 square miles in size, which is much smaller than previously thought.[19]

Wild boar rub against trees to mark their territorial boundaries. After wallowing in mud, they cover these "rubbing trees" with their scent. Scent markings, however, are not permanent, which means that boundaries between territories remain somewhat fluid, and so it's little wonder that every once in a while, boar intrude where they don't belong. As meeting up with strangers usually leads to altercations that even boar prefer to avoid, border violations by unrelated sounders are fairly rare, but if the home ranges of two related groups are side by side, their territories may overlap by as much as 50 percent. Clearly, wild boar are more kindly disposed to family members than they are to strangers and, most importantly, they can obviously tell the difference.

Family dispersal starts when the previous year's piglets, the yearlings, are driven off as the birth of the next litter approaches. The sow has no extra time to look after older piglets, which are pretty independent by then. Wild boar are highly social and love to engage in mutual grooming or simply to lie snuggled closely together, and the siblings join up to form yearling sounders so they can continue living in a group. If the yearling sounders run into their extended families with their new piglets later in the year, the meeting will be a friendly one. Everyone knows everyone else and they all still get along well.

Thinking about our domestic animals, I've often wondered whether our goats and rabbits are capable of picking their grown-up children out of the group and recognizing them as relations. After observing them for a long time, I believe I can now answer this question with a resounding yes. With one proviso: that the animals are not separated from one another. If they are kept in separate enclosures for more than a few days, they end up treating each other as strangers. Perhaps their long-term memory is not configured for storing information about family relationships. It is clearly different for wild boar, and therefore also probably for domestic pigs, because they have long memories for who belongs to whom. This is of little use to domestic pigs, of course, because unfortunately for them, they are separated from their parents and raised in groups of other pigs their own age, and as a rule, they don't make it past their first year.

These days, most people are aware that pigs are extremely clean animals. They prefer to use some kind of a toilet—a designated place where they do their business—be it big or small. This toilet is never in their sleeping hollow. After all, who would want to sleep in a stinky bed? This goes for both wild and domestic pigs. When you see photographs of tiny stalls in factory farm barns (11 square feet per animal) and pigs covered in muck, you can imagine how uncomfortable the animals must be. In the wild, boar adapt their sleeping quarters according to the weather and time of year. Because

39

they select their resting places with care, they prefer to use the same spot all the time; however, when storms blow and rains soak their slumbers, they will move to wooded areas where they can sleep protected from the wind and stay relatively dry. In summer, the bare earth is sufficient as a place to lay their head, because at that time of year wild boar are usually too warm anyway. In winter, however, they plan their nightly repose especially carefully. A snug little nest under a thick windbreak of blackberry bushes with only two or three tunnel-like entrances is ideal. They bring in dried grass and leaves, moss, and other soft materials, which they pile up carefully to make a cozy bed.

Did I say "nightly repose"? Although they would probably love to sleep as we do when we lie dreaming in our beds, these smart animals have adjusted their circadian rhythms. Every year in Germany, hunters shoot as many as 650,000 wild boar,[20] and to do that they need daylight. In order to avoid their pursuers, the boar go about their business under the cover of darkness. Normally, that would be protection enough, because in Germany it is illegal to shoot animals after dark; however, an exception has been made for wild boar to try to control their burgeoning populations. Because night-vision devices are still prohibited, hunters have to wait for a full moon and clear skies so they can see more than just vague shadows in woodland clearings. They attract the wild boar with small amounts of feed corn, which the boar are particularly partial to. The goal: dispatch the boar with a deadly shot while they are feeding. But it's not that simple to outwit the canny animals, which simply put off feeding until the wee hours of the morning. But the hunting industry has a solution at the ready for this, as well: game clocks. These are clocks that stop when they are disturbed. When hunters put these clocks in among the corn, they show the time when the wild boar come to feed. Now hunters can climb up into their blinds at just this time, and they don't have long to wait until their prey appear.

In the final tally, however, it is the wild boar that seem to have come out ahead. In some cases, they rely on the bait as their main source of food, and they multiply quickly despite the hunting pressure—so much so that reducing populations has become a lost cause in many places. In the US, feral hogs are generally diurnal if left undisturbed, but they react like their European counterparts and intensive human activity or hunting during the day can drive them to be active at night.

There have been many particularly touching discoveries made about domestic pigs, if only because a number of research facilities are working to improve factory farming. When the newspaper *Die Welt* asked Professor Johannes Baumgartner at the University of Veterinary Medicine in Vienna if there had been any notable characters among the pigs he had studied, he told them about one old sow. Over the course of her life, she gave birth to 160 piglets. She taught all of them how to build a nest out of straw, and when her daughters grew up, the old sow assumed the role of midwife and helped them prepare for the births of their own babies.[21]

The question then becomes, if researchers know so much about the intelligence of pigs, why isn't the image of the smart pig publicized more? I suspect it has to do with eating pork. If people knew what kind of an animal they had on their plate, many would completely lose their appetite. We already know this from primates: could any of us eat an ape?

6.

GRATITUDE

I T SHOULD BE clear by now that whether they are driven by their circumstances or our desires, whether they want to or not, animals love people (and, of course, the reverse is true). I consider gratitude to be a closely related emotion. And animals can certainly feel gratitude, as well. Owners of dogs with checkered pasts that have been welcomed into families later in life are particularly well placed to confirm this.

Our cocker spaniel, Barry, didn't come to us until he was nine years old. Actually, after the death of our Münsterländer, Maxi, we wanted to draw the dog chapter of our lives to a close. Or so we thought. Despite the fact that my wife, Miriam, was absolutely opposed to a new family member, our daughter set out to convince us otherwise. She didn't get much resistance from me, because I couldn't really imagine life without a dog. When my daughter accompanied me to a fall market at a nearby country store, both of us were aware of what might happen. The Euskirchen animal shelter was going to have a parade of its guests, hoping to find homes for them that day. My daughter and I were hugely disappointed when the only animals on display were rabbits, because we already had plenty of those at home. After waiting around the market all day, making multiple tours of the stalls, here we were, confronted with this—no dogs. Right at the very end, there was an announcement that a future occupant was going to be shown by one of its former owners before being delivered to the shelter: Barry. Our hearts beat faster. The dog was apparently extremely good natured, a model passenger in the car, and he was neutered. Perfect! We leapt up from the bench and stepped forward. A short test walk, a handshake to seal the agreement for three days' probation, and we took off right away with the dog in the car, headed for Hümmel.

The three-day trial was important, because Miriam didn't suspect anything yet. She came back late that night after an engagement. She was taking her coat off, when my daughter asked: "Do you notice anything different?" My wife looked around and shook her head. "Then take a look down at your feet," I prompted. And in that instant, it happened. Barry looked up at her, wagging his tail, and my wife took him into her heart right then and there for the rest of his life. And the dog was grateful—grateful that his long and arduous journey had finally ended. His owner, an old lady suffering from dementia, had had to give him up. He'd gone through two different families, and now he had found his forever home with us. It's true that for the rest of his life, he worried that there might be yet another handover, but other than that Barry was always happy and friendly. He was grateful. It was as simple as that—or was it?

After all, how are you supposed to measure gratitude or—what's almost as difficult—to define it? If you check on the web, you'll find a lot of discussion but nothing definitive. Some animal lovers think of gratitude as their due, a response many owners expect from their animals in return for the care they give them. I wouldn't even bother to search for this kind of gratitude in animals, for it would merely be an expression of subservience smacking of servility. Essentially, and this is in reference to people, what emerges from most definitions is that gratitude is a positive emotion arising from an enjoyable experience caused by someone or something else. In order to be grateful, you need to be able to recognize that someone (or life) has done you a good turn.

The Roman politician and philosopher Cicero considered gratitude to be the greatest of all virtues, and he thought dogs were capable of feeling it. But now it gets tricky. How can I know whether an animal recognizes who or what has caused its enjoyable experience? In contrast to the joy itself (which is easy to recognize in a dog), there's also the question of whether the dog gives any thought to the cause of its joy. It's relatively simple to answer this question. There's food, for starters. The dog is happy about its meal and knows exactly who filled its bowl. In fact, dogs often encourage their owner to repeat the process. But is this really gratitude? You could just as easily call it begging. Doesn't true gratitude include a mindset, a way of looking at life? An ability to celebrate small pleasures without constantly craving more? Seen from this perspective, gratitude is when joy and contentment about circumstances that are not of your own making coincide. Unfortunately, this kind of gratitude cannot yet be proven in animals—we can do no more than speculate about their inner outlook on life. In Barry's case, at least, my family and I are certain that he was both happy and content to have found his final home with us, even if we don't have any scientific proof.

But how about other examples in the animal world? Might they shed more light on the issue? There is the story of a humpback whale in the Sea of Cortez in Mexico that put on an hour-long display of breaching and flipper flapping after a man called Michael Fishbach spent hours cutting off a fishing net in which it had been hopelessly entangled. When Fishbach encountered the whale, it looked as though it would not be able to survive for much longer. Fishbach immediately entered the water armed with just a small knife. As soon as the whale was free, it put on a glorious acrobatic display. Perhaps it was just happy not to be entangled in the net anymore—or perhaps it was performing for the people in the small boat to thank them for rescuing it from certain death.[22]

Then there is this story of gratitude closer to home about wild birds that voluntarily shared their treasures with a young girl in Seattle. Crows like to collect shiny things and, of course, crows also like to eat. When Gabi was four, like many four-year-olds, she was not a particularly neat eater, and sometimes she dropped her food on the ground when she was outside. The crows lost no time snatching it up. A couple of years later, Gabi began intentionally sharing her lunch with the crows as she walked to the school bus stop. Then she began feeding the crows on a daily basis in her backyard. Soon after that, the birds began bringing her gifts: bits of glass, screws, pieces of bone, broken jewelry. One time Gabi's mother, Lisa, dropped the lens cap of her camera when she was photographing the crows and thought it was lost, only to find it right there a few days later on the bird feeder. The crows had even carefully washed it before returning it. Is this gratitude, perhaps? Crows are certainly known for recognizing people and for having strong emotional reactions to those they don't like. In this case, it seems they liked Gabi very much, and perhaps they were grateful and thanking her for looking after them for all those years.[23]

7

LIES AND DECEPTION

CAN ANIMALS LIE? If you define the term loosely, then quite a few can. The hoverfly, whose yellow and black stripes make it look like a wasp, "lies" to its enemies by making them believe it is dangerous. It must be said that the fly is unaware of its deception, because it doesn't actively undertake it; it was just born looking that way. It's the same with the European peacock butterfly. With big "eyes" on its wings, it signals to its enemies that it's bigger than it really is and is too large for them to tackle. But let's put these examples of passive lying aside and take a look to see which animals are the real tricksters with deception on their minds.

One of these, for example, would be our rooster, Fridolin. He's a portly representative of his species and as white as driven snow, which is just as he should be because he's a white Australorp. Fridolin lives with two hens in a nearly

200-square-yard run designed to keep out foxes and hawks. Two hens are quite enough to provide us with the eggs that we need. Fridolin, however, sees things differently. Such a small flock doesn't keep him as busy as he'd like to be, and with a sex drive like his, he could easily satisfy a couple of dozen lovers. Circumstances, however, dictate that he has to concentrate all his loving on Lotta and Polly. The hens are not at all keen on constant coupling and therefore quickly give Fridolin the slip as he prepares to make that final pounce. If he somehow manages to land on the back of one of his ladies despite her best efforts to avoid him, he spreads his wings to keep his balance. At the same time, he grabs hold of the neck feathers of the hen, which is now squashed down flat on the ground, and sometimes he even pulls feathers out in a frenzy of passion. Then he presses his cloaca against hers and squirts his semen inside. As soon as the act is over—it lasts just a matter of seconds—the hen gets up and gives herself a shake and, at least for the time being, she can return to eating un-molested. But it's not long before Fridolin's ardor returns, and because neither hen is interested in joining in, he has to chase them down all over again. The rooster often runs out of steam while pursuing them, and a modicum of peace descends.

But then Fridolin came up with an easier way of getting what he wanted. Fridolin is usually a real gentleman, allowing his little harem first dibs at food. As soon as he spies some-thing tasty, he makes a special clucking sound, and it's not long before Lotta and Polly rush over and fall upon the food he has found for them. But sometimes there isn't anything to eat under Fridolin's feet, and it turns out that the rooster has enticed his hens over with a bald-faced lie. Instead of tasty worms or crunchy seeds, what awaits them is another of Fridolin's attempts to mate, and—thanks to the element of surprise—his devious plan is often rewarded with success. However, if he tries this trick too often (and with two hens a couple of lies suffice), then both of them become cautious, even when real food is on offer. No one believes a liar, even when he tells the truth ...

Other species of birds can also be big fibbers. Take swallows, for example. If a male returns to find his mate is not on the nest, he gives an alarm call. His mate wrongly supposes danger threatens, and she flies back to the nest via the shortest route. The male uses the false alarm call to keep his mate from pursuing dalliances in his absence. Once the eggs have been laid, this is no longer a concern, and the deceitful calls disappear.[24]

Great tits provide another example. They are widespread in many parts of the world and a number of them are fibbers, for when it comes to food, it's every bird for itself. These pretty birds with black heads and white cheeks have a sophisticated language they use to warn each other about predators. One of these predators is a small raptor called a sparrowhawk that likes to hunt in backyards. The bird dives down, quick as an arrow, to grab sparrows, robins, and tits, and then flies to the nearest bush to eat them. A great tit that spies the danger from afar warns others of its kind with a high-pitched call. The call is out of the sparrowhawk's range of hearing, which gives the great tits a chance to slip safely away before the raptor spies them. If, however, the raptor is already dangerously close, the warning is broadcast at lower frequencies, so all the tits know an attack is imminent. The attacker can also hear this deep churring call and immediately realizes that its planned surprise attack isn't a surprise anymore. It often ends up grabbing thin air because the tits are on the alert. I'm sorry to report that some great tits take advantage of their well-functioning community. If there's particularly tasty food around or if food is scarce, then the little liars make the usual alarm call. All the birds quickly fly to safety—or almost all. Left alone, the trickster can now eat as much as it likes.

How about cheating on your mate? This form of sexual liaison is also a kind of deception—at least it is when the cheater knows what it is doing. And you can see exactly this with male magpies. Some urban communities consider these handsome black-and-white corvids to be public enemy number one, because they snatch the babies of other songbirds to feed their own young. This behavior puts them in the same league as the squirrels I've already talked about. I like to imagine what it would be like if magpies were an endangered species. If that were the case, we would be excited to see them, and we would marvel at how their black feathers shimmer blue-green in the light. Unfortunately, people who appreciate their beauty are few and far between.

But back to cheating. Like other corvids, magpies form pair bonds that last a lifetime. They set themselves up with their partners in a home territory that, like their partnership, lasts for many years. They vigorously defend their home from others of their kind. This is clearly because both partners want to avoid sexual dalliances, because after the eggs have been laid and most of the business of procreation has been accomplished, the zeal with which they defend the boundaries of their territory diminishes considerably. But even before that happens, much of this territorial defense is strictly for show, at least on the part of the males. While the female aggressively drives off any competitors that intrude, her partner is an opportunist. As long as his mate is watching or within earshot, he'll be just as aggressive in driving off any intruding female. However, if he thinks he's not being observed, he begins eagerly courting the attractive stranger instead.[25]

There are, however, other strategies in the animal world that cannot really be called lies, even if this is how they are sometimes described in the media. There are reports of foxes that, in contrast to peacock butterflies, actively fool others. The fox's hunting repertoire includes playing dead, and a fox may even let its tongue hang out to make the scene more convincing. A corpse out in the open? There are always takers, most often crows, which are happy to help themselves if there's delicious meat on offer, even when it's a bit past its prime. In the case of the fox, it's super fresh—too fresh, as it turns out. Any black-feathered guest that decides to grab a bite suddenly finds itself in the jaws of the wily fox and ends up being the meal instead.[26] That is a masterful piece of play acting and definitely a trap, but it is far from being a lie. Lying usually involves deceiving others of your own kind by providing them with false information in order to get an advantage for yourself. The fox is simply following a hunting strategy that is cunning and morally above reproach. The fox is quite different from Fridolin or from the magpie that sneaks a bit on the side, both of which are intentionally pulling a fast one on fellow animals near and dear to them.

No matter how you might judge these cases of trickery from a moral standpoint, I, for one, am moved by them because they show just how intricate the inner lives of animals really are.

8

STOP, THIEF!

I F LYING IS widespread among animals, then what about thievery? A good place to start is by looking at socially oriented animals, because thievery, like lying, has a moral component, and it is judged to be negative only when it is socially relevant and when it adversely impacts other animals of the same species.

When it comes to thievery, the American gray squirrel is a sly one, but before delving into its behavior, let's check out how it's doing in Europe, where it has become a serious threat to native red squirrels (which can also be black in color). In 1876, a certain Mr. Brocklehurst from Cheshire, England, took pity on a pair of grays that were being held captive, and he released them. In the years that followed, many dozens of other animal lovers followed his example. The gray squirrels thanked their liberators by reproducing enthusiastically—so enthusiastically that they have now pushed their red European relatives almost to the brink of extinction. Gray squirrels are larger and more robust than their red cousins, and they adapt to any kind of woodland, whether deciduous or coniferous. Even more dangerous for native European squirrels, however, is a stowaway that hitched a ride along with the grays: squirrel poxvirus. Whereas American grays are generally immune to the virus, European red squirrels die in large numbers when they are infected. Unfortunately, not only were grays released in the UK, they were also released in northern Italy in 1948, and since then they have been advancing on the Alps. We don't know if at some point they'll manage to scale the mountains and march victorious into the German woods, as well.

Despite all this, I don't want to characterize gray squirrels as vermin. After all, it's not their fault they were brought to Europe, and their dominance is not due to their behavior, which brings us back to the topic of thievery. Squirrels sometimes get food by plundering the winter caches of other squirrels. In many cases, they must do this to survive, as the unsuccessful searches in the snow that I observe from my office window every winter attest. A squirrel that cannot remember the location of its caches will starve, and as a last resort, it may help itself to its neighbors' loot. I don't know whether European red squirrels have developed a counterstrategy, but researchers discovered that American grays have. A team from Wilkes University in Philadelphia observed gray squirrels digging empty caches. The squirrels did this in clear sight of others of their kind to lead them astray, but only when they thought they were being watched. When the eyes of other squirrels were on them, they dug around a bit in the dirt and pretended to bury something. According to the researchers, this is the first report of rodents using deceptive tactics. When lots of unfamiliar squirrels were watching, up to 20 percent of the caches were empty. As part of the experiment, the researchers then had students raid the caches that had food in them, and guess what happened? The squirrels reacted immediately and from then on used deceptive tactics in the presence of suspect humans, as well.[27]

Thievery is also a big deal for Eurasian jays. These birds are fanatical about survival. In the fall, each bird caches up to eleven thousand acorns or beechnuts in the soft forest floor, even though they could survive the winter just fine with far less food. Not only do they rely on the oil-rich seeds as emergency rations until the next growing season, but they also feed them to their chicks in spring. Even allowing for this, the sly birds usually store way too many seeds. And what an amazing memory they have: jays find every one of their thousands of caches with a single stab of their beak. Small trees sprout from the unused seeds to ensure that future generations will have their own supply of nuts and acorns.

In the woodland I manage, we use the birds' passion for collecting to plant young deciduous trees in the monocultures of old spruce plantations. This is how it works. We put seed trays on posts and fill them with acorns and beechnuts. Jays love to come and help themselves, and they distribute their booty in the soil hundreds of yards in every direction. It's a win-win situation. We get precious new stands of deciduous trees in the woodland, and the jays get huge quantities of winter provisions with very little effort. Some years, however, the oaks and beeches do not set seed, and then things get tight for these colorful birds. Whereas the population increases in years of plenty, in lean years it shrinks. This ruthless natural cycle has been repeated over and over since time immemorial. But who wants to starve? Some of the birds fly south, while most try to survive in the woodlands they call home.

Just like the squirrels, in lean times jays watch other jays in late fall to see where they bury their treasures. And because no bird can keep watch over such a large number of hiding places, sneaky individuals can live well over the winter by profiting from the hard work of others. Scientists at the University of Cambridge have discovered that the birds are well aware of these shenanigans. They discovered this by putting trays filled with two different materials into the aviaries. Some trays contained sand; others contained gravel. Whereas sand doesn't make any noise when you dig in it, gravel gives the game away by rattling. And the jays kept this in mind when they buried their caches.

If the jays were alone in their enclosures, it didn't matter to them whether they hid the proffered peanuts in sand or in gravel. If the competition could see and hear them when they were digging, it also didn't matter which material they were rummaging around in. In the first case, no other bird was around to witness where the precious booty was hidden. In the second, the birds realized that any bird watching them would know where the food was anyway. However, if the competition was out of sight but still within earshot, the jays opted for the quieter sand. Under those conditions, there was a much higher likelihood that the potential thief would have no idea anything had been hidden. And for their part, the thieves were also quieter. Whereas they normally called loudly when they saw other jays, when they were watching food being hidden, they were considerably less vocal, clearly to avoid betraying their presence.[28] This experiment clarified two things. First, the bird doing the hiding can put itself in the position of other jays and take into account what they can and cannot see. And second, the future thief was obviously planning its actions in advance, because it limited the sounds it made in order to increase its chances of later plundering the cache of peanuts undisturbed.

Of course, theft in the sense of the intentional seizure of assets that don't belong to you doesn't occur only within a species. Come winter, in many deciduous woods you can find traces of interspecies plundering. Sometimes you come across holes in the forest floor one to two feet deep, their edges strewn with big clumps of excavated earth. Wild boar are the only animals that root around like this, and they do it in so-called mast years. This technical term describes years when beeches and oaks go into overdrive to produce seeds. These years were once a blessing for farmers. A mast year meant they could drive their domestic pigs into the woods one last time to fatten them up before the winter slaughter. Farmers are not allowed to pasture their animals in woodlands any longer (at least, not in Central Europe), but the term "mast" (from the German *mästen,* "to fatten") is still used.

Naturally, wild boar behave the same way as their domestic relatives in a mast year: they put on a nice thick layer of fat. But once the unexpected gift has been cleaned up and all the nuts lying on the ground have been scarfed down, rumbling stomachs demand a top-up. And snacks can still be found deep down in the soil. This is where mice have buried stores—pantries stocked with their portion of the harvest so they can make it safely through the winter. Even in times of hard frost, the ground doesn't freeze more than a few inches below the insulating layer of leaf litter, and in the mouse's quarters, it's always at least 41 degrees Fahrenheit. Thanks to a cozy layer of leaves and moss and a completely draft-free location, a mouse can survive very well down here. That is, as long as no wild boar come calling.

The gray hustlers have extremely sensitive noses and can smell the mouse's digs from yards away. From experience, they know that the little creatures diligently collect beechnuts and other seeds, and store them all together in one convenient spot. An enormous store that will last a mouse for months is a tiny snack that will tide a wild boar over until its next meal. However, since mice mostly live in large colonies, a number of small snacks can provide the calories a wild boar needs to make it through a cold winter's day. And so the boar burrow along the underground tunnels, smashing storerooms and emptying them in a couple of gulps. The only option the mice have is to flee and face an uncertain future, for in winter there are very few sources of food for the homeless. If they can't evade the wild boar underground, they are gobbled up along with their food stores; boar enjoy their veggies with meat on the side. At least the swallowed mice are spared a long, slow death from starvation.

And what does this behavior look like from an ethical point of view? The wild boar's plundering of mouse assets isn't really theft, because they are not deceiving other boar. They are perfectly aware they are raiding mouse provisions, but this is a completely normal way for them to get food, even if the proceedings look very different from the perspective of the mice.

9

TAKE COURAGE!

I F ANIMALS FUNCTIONED only according to fixed genetic programming, then each individual of one species would react the same way under the same circumstances. A certain amount of a hormone would be released that would trigger the corresponding instinctive behavior. But that is not the case, as you probably already know from observing domestic animals. There are courageous and cowardly dogs, aggressive and super gentle cats, jumpy and bombproof horses. The character each animal develops depends on its individual genetic predisposition and, just as importantly, on the influence of its environment, which is to say its life experience.

Our dog Barry was a little scaredy cat. As I have already mentioned, before he came to us he had already been passed along by a number of different owners. For the rest of his life, he was scared of being abandoned, and he always got extremely worked up when he was taken along when we visited friends. If you are a dog, how are you supposed to know whether you're going to be handed off yet again? He showed his nervousness by panting non-stop, so we finally gave up, leaving the distressed animal alone in the house for a couple of hours instead. When we got back, it was easy to check whether or not Barry was relaxed. He became deaf in his old age and couldn't hear us arrive, sleeping soundly until he blinked up at us when he felt the wooden floorboards vibrate under our feet. So Barry is an example of an animal that lacks courage, but we wanted to take a look at the opposite trait, and to do that, let's step out into the woods.

One fawn that had breached a plantation fence along with its mother showed particular courage. I used to erect these fences around areas where storms had toppled trees in monocultures of plantation spruce. In order to allow as natural a woodland as possible to regenerate, forestry workers planted little deciduous trees. These newly planted areas needed to be protected from the greedy mouths of browsers, and that's why I erected the fences. The wire fences behind which the oak and beech saplings grew were six feet tall. During a late-season storm, a spruce nearby had fallen on one of these fences, flattening it. Deer, including the aforementioned doe and her fawn, had wandered through the gap directly into a land of ease and plenty. No walkers disturbed them there, and they could munch away on the tasty shoots of much sought-after deciduous trees. Things looked a little different to me. The expensive fence was no longer of any use, and the goal to have a halfway natural beech and oak wood one day was fast disappearing. And so, accompanied by my Münsterländer, Maxi, I climbed in after them to drive the freeloaders back out.

To do this, I made an opening in one corner of the fence so that when I drove the deer along the inside of the wire, they could escape. Once Maxi got in on the action, this was the only way they could go. She responded to my hand signals from about a hundred yards away, dashing here and there, flushing the intruders out of the shrubby undergrowth. One of the deer that had found its way in rushed passed me and out of the opening, but then, 20 yards farther along, it crawled back in on its stomach through a tiny gap under the fence. Things weren't working out with the doe and her fawn, either, on account of the fawn. The mother was now trying to lead it out at a gallop, and Maxi was chasing it at full tilt in the direction it needed to go, when suddenly, it all got too much for the youngster. It turned around and rushed menacingly toward the dog. Maxi was normally very courageous and there wasn't much that frightened her, but a fawn coming right at her? She had never experienced anything like it. She pulled up, baffled, but the fawn continued its charge. Maxi finally turned and fled. That was it for me that day, and the deer were allowed to stay in the newly planted area. They had completely lost respect for the dog, and all I could do was shake my head and laugh. I'd never encountered such courage in a young fawn. And it had shown courage, because its mother should have been the one to step in and defend her offspring.

But what exactly is courage? Once again, this term has a variety of vague definitions (I invite you to try to come up with a definition off the top of your head), although one general concept seems clear: courage involves realizing that it is important to act despite recognized danger and then doing so. In contrast to bravado, courage is considered to be a positive quality, and in this sense of the word, the fawn had certainly acquitted itself well.

Just as courageous, by the way, are the fieldfares I mentioned earlier that raise their broods in the old pines by our forest lodge. When a carrion crow, their old archenemy, shows up, they don't just sit there and watch it grab their chicks. As soon as the feared bird starts to fly toward the colony, an aerial attack is mounted. The little thrushes swarm around the considerably larger intruder and dive-bomb it incessantly. It would be easy for the crow to fight off or injure the tiny raging birds, but their determined preemptive strikes, usually mounted in concert with other thrushes, throw the crow off, and it starts weaving to evade its attackers. It doesn't notice that the attacks are driving it away from the nest (which, of course, is what the thrushes want), and it also seems to find the attacks really annoying, because after a few minutes, the crow usually beats a retreat and disappears from the stand of old trees. Does that mean that fieldfares are courageous? Or are they simply running a genetic program activated by the appearance of their enemy? It's a mixture of both, and that's just how it is in every similar situation—probably even with us. Not all thrushes react in such a spirited fashion or, above all, with such persistence. How far they pursue the crow and how fiercely they dive-bomb it is different with every bird. And while fearful thrushes mount only half-hearted attacks, the courageous masterfully manage to drive the crow off for several hundred yards.

But are the less brave birds automatically at a disadvantage? Niels Dingemanse and his team at the Max Planck Institute for Ornithology don't think so. They investigated great tits for similar character traits and discovered that the shyer individuals get on better with other great tits. They avoid disputes and large flocks and prefer to live in small groups of like-minded birds. Shy birds are slower and calmer, and take their time before mobilizing. And this means they notice things their courageous, speedy colleagues often overlook, such as seeds left over from the previous summer.[29] Because the advantages and disadvantages to being either courageous or shy seem to balance out, both character traits have survived until today.

10

BLACK AND WHITE

MANY PEOPLE ARE interested in animal emotions in principle, but their interest doesn't usually include all species, particularly not those considered dangerous or disgusting. People often ask me: "What use are ticks?" The question amazes me to this day, because I do not believe that any one animal has a more important job to do in an ecosystem than another. That might sound odd coming from a forester, but I think a principle like this extends to all creatures the respect they deserve.

But let's take this one step at a time. First, let's look at more insect examples—wasps, for instance. In late summer, these social insects can get really annoying, and at some point even I have had enough of these stripy little stingers. Perhaps this goes back to an experience I had in my youth. I was racing along on my bicycle on the way to the swimming pool, when a wasp flew at me and, because the wind was in my face, it got stuck between my lips. I pressed my mouth shut, but I could do nothing to stop the wasp from stinging me liberally. I felt as though my mouth was being sewn shut. My lower lip imme-diately swelled up so much I was afraid it was going to burst. Then there was the fact that at that age you tend not to be very self-confident when it comes to physical blemishes. Since then, I've not been very keen on wasps. You might have had a similar experience yourself, and so it's hardly surprising that you can buy all kinds of gadgets to keep them away. There are those bell-shaped glass contraptions that you fill with entic-ing sugary liquids to attract the wasps and then drown them. It sounds mean, and it is. But stinging insects are usually not held in high regard, and we're happy to dispatch them quickly.

Let's go ahead and move on to a row of cabbages in a raised bed belonging to a colleague of mine. A whole bunch of fat cabbage white caterpillars were sitting on the succulent leaves. Gardeners categorize these caterpillars as pests, because they riddle cabbage leaves with holes all the way down to the ribs. My colleague asked us for advice and we were able to help her out: we had had good experiences with neem oil for a number of years. Once we started using this ecologically friendly spray (which is also certified for use in organic agriculture), our cab-bages made it through to harvest intact. But we're not using neem oil anymore, and this is where the wasps come in. They pounce on the caterpillars and munch them into little pieces so they can carry their bite-sized booty back to their nests to feed their hungry brood. Just like that, all the caterpillars disappear, as we can attest from our experience at our forest lodge. A plague of wasps in the summer means caterpillar-free cabbages. Does that mean wasps are beneficial?

Most of the creatures in our garden have earned labels like this. Tits: beneficial (eat caterpillars). Hedgehogs: beneficial (eat snails). Snails: pests (eat greens). Aphids: pests (suck the juices out of plants). How wonderful that for every creature that is considered a pest there is a beneficial one to keep it in check. But if you divide Nature like this, you automatically assume two things. First, there must be a Creator who has planned and implemented a design where everything is finely tuned and in balance. Second, this Creator has made our world in such a way that it is completely oriented to our needs. Questioning the point of ticks makes sense in this world view. I don't want to criticize. After all, this way of thinking is widespread even among conservation groups that encourage beneficial creatures, for instance, by building nest boxes for them. But can Nature really be neatly compartmentalized like this? And if it could, where would we fit?

No, I believe that the unimaginably lively lives of millions of species have adapted so well to one another only because overly selfish species end up recklessly exploiting all available resources, destabilizing ecosystems, and irrevocably changing them and their inhabitants. Just such an event played out about 2.5 billion years ago. At that time, many species lived anaerobically, which is to say without oxygen. For life at that time, the most important gas we breathe today was pure poison. One day, cyanobacteria began to spread with astonishing speed. They got their food by photosynthesizing, and as they did so, they released a waste product into the air: oxygen. At first, this gas was taken up by rock, and rock that contained iron, for example, rusted. But at some point, there was so much excess oxygen that the air became increasingly oxygen rich until finally, a deadly threshold was crossed. Many species died out, and the ones that remained learned to live with oxygen. At the end of the day, we are the descendants of the creatures that adapted.

Basically, small adaptations are being made every day. What we understand as a finely tuned balance between prey and predators is in reality a harsh struggle with many losers. When a lynx is wandering through its enormous territory, it's hungry for deer. However, the cat is not a good sprinter and so it must rely on the element of surprise. Clueless, careless browsers that have not yet heard the news about a large predatory cat in the area are particularly easy prey. A lynx can enjoy a deer a week, but only until all the deer in the area have been tipped off to its presence. Then, panic prevails at the snap of a twig and even pets are wary. A colleague of mine tells me that his cat is the first to alert him when there's a lynx in the area. The cat refuses to set foot outside, but he can't say what alerts the cat. Perhaps it is the collective behavior of potential prey that gives rise to an aura of suspicion in the wood. Because of this, the lynx makes fewer kills and is forced to move on. It is only a few miles farther on, in a new area of unsuspecting prey, that it can get back to easy pickings. If there are too many lynx on the prowl in the same area, at some point there will be no more unsuspecting prey. Then, when winter ushers in lower temperatures and a corresponding rise in energy needs, many lynx starve—especially the younger, less experienced animals. You could say the population is self-regulating, but what that really means is that animals die ghastly deaths.

So Nature is nothing like a neat set of compartments. No species are inherently good or bad, as we have already seen in the case of squirrels. But it is much easier for us to empathize with or at least take an interest in squirrels than it is for us to relate to the ticks mentioned at the beginning of this chapter. And yet even these abhorrent little creatures have feelings. The presence of something as simple as hunger is empirical proof of this, for the tiny arachnids go after mammal blood only when their stomachs are rumbling. And this means an empty stomach must feel unpleasant, particularly when it hasn't been filled in almost a year, which is how long ticks can hold out until their next meal if they absolutely have to.

When a large animal crashes by, ticks can feel the disturbance and smell sweat and other animal odors. They quickly extend their forelegs, and with any luck, they grab onto the legs or body brushing by and hitch a ride. Eventually, the ticks crawl to a comfortable, warm spot where the skin is not too thick, and there they settle in. They push their extended mouthparts into the wound and slurp up the blood that seeps out. The tiny vampires can increase their weight many times when feeding, and they swell up like little peas. On the way to maturity, they must undergo three molts, and before each stage they need to find a new victim to fill up on. That's why it can take up to two years for them to reach adulthood. When it gets to the stage when the smaller males and the larger females have sucked themselves full to bursting, all that remains is the finale.

The little males must mate. Must? They want to. Just like us, they are driven by desires, and they eagerly seek a partner they can grab hold of to complete the final act. After that—and here, I hope, is where the similarities end—they die. The females live long enough to lay up to two thousand eggs, and then they, too, pass away. The tick is a creature whose ultimate joy, or—because this cannot yet be proven—at the very least the high point of its life, consists in producing thousands of offspring and then dying, completely spent. If it were a mammal, we would call this behavior self-sacrifice. For ticks, unfortunately, people currently reserve just one emotion—disgust.

11

COLD HEDGEHOGS,
WARM HONEY BEES

W E ALL LEARNED this in our biology lessons in school. Apart from all the other divisions you can make, the animal world can also be divided into warm-blooded and cold-blooded creatures. Yet more compartments and, as you will see, animals don't fit neatly into these, either. But first let's get to the scientific categories. Warm-blooded animals regulate their own body temperature and keep it constant. Humans are a good example. When we get cold, we shiver to generate the heat we need. When we overheat, we sweat and cool down as our sweat evaporates. In contrast, for better or worse, cold-blooded animals are dependent on external temperatures. When it gets too cold, they can forget about physical exertion. And that's why, every winter, I find flies in my wood pile that just don't have the energy to get up and go. When temperatures drop below freezing, they are not up for much more than crawling around excruciatingly slowly among the logs. Helpless, the most they can hope for is that no birds track them down when it's cold. And that's how it is for all insects. All of them? Actually, no, that's not how it is for my bees (or any other bees, for that matter).

I used to not like bees. It's difficult to forge a relationship with insects, and if those insects sting, the almost automatic response is aversion. Apart from that, I rarely eat honey. Not an auspicious start for a beekeeper, which is what I've become. It's really only because of the apples. We had hardly any bees in our orchards in spring. To change that, I acquired two hives in 2011. Since then, pollination has not been a problem, and we've had plenty of honey. But, most of all, I've learned that there are ways in which bees are not the same as other insects. For example, in some respects they are similar to warm-blooded animals, and that is the main reason they are so eager to forage. Nectar, processed into honey and stored in combs, serves as a winter fuel for bees, for they love to be cozy and warm. Their comfort zone lies between about 91 and 96 degrees Fahrenheit, just a little cooler than the body temperature of mammals.

In summer, that's not a problem. Quite the opposite, in fact. The muscle action of up to fifty thousand busy bees creates a decent amount of heat, which must be dissipated so the hive doesn't get too hot. This is a complicated task. Worker bees carry in water from the closest source so it can evaporate and cool the interior of the hive. Thousands of wingbeats circulate air in the hive, creating a cool breeze between the honeycombs. Too much disturbance overwhelms this communal effort. If the hive is attacked from the outside or if it is handled incorrectly when it is being transported from one place to another, the agitated bees heat up so much that the combs melt and the insects die of heatstroke. The technical German term for this translates as "death by buzzing." The expression comes from the noise the colony makes as the bees panic and beat their wings faster, heating up the hive and causing their own demise.

In the normal course of events, however, the bees' thermoregulation processes work perfectly. For my hives, most of the year it's too cold rather than too hot, and creating heat is most important. Vibrating muscles means calories spent, and bees take in the energy they need in the form of honey. Honey is basically a thick, highly concentrated sugar solution with vitamins and enzymes added. Every month, primarily in winter, my bees burn through more than 6 pounds of honey per colony. Honey serves the same function for bees as fat does for overwintering bears, and just as bears emerge from hibernation much skinnier than they were in the fall, so the size of colonies can shrink tremendously in cold weather. If it gets really cold, the insects huddle together and form a ball. It's warmest and therefore safest in the middle—and, of course, this is where the queen must be. And what about the bees on the outside? If the exterior temperature drops below 50 degrees Fahrenheit, they would die of cold in just a few hours, so bees inside the ball are kind enough to spell them off to give them the opportunity to warm up again in the dense, seething mass.

Bees prove that not all insects are exactly cold-blooded, and as you've probably guessed, not all mammals are exactly warm-blooded, either. Maintaining a constant body temperature is supposedly a mammalian (and avian) specialty. Supposedly. The little hedgehog is the exception that proves the rule. While the similarly sized squirrel bounces through the branches every once in a while in the cold season, even in the snow, this spiny ground dweller spends its winters fast asleep. Its spines don't insulate as well as a squirrel's thick fur, and therefore it uses a great deal of energy when temperatures fall. Apart from that, its favorite foods—beetles and snails—have made themselves scarce and are no longer to be found above ground. So what better idea than to take a break as well? The prickly little guys roll up into a comfortable ball in a cozily padded nest that is often buried deep beneath a pile of leaves or brush. Here, they fall into a deep sleep that can last for months. In contrast to many other mammals, instead of keeping their body temperature at a hedgehog-appropriate 95 degrees Fahrenheit, they simply shut off their energy intake, which means their body temperature falls to match the ambient temperature and sometimes drops as low as 41 degrees. Their heartbeat slows from up to two hundred beats to only nine beats per minute, and they breathe only four times a minute instead of fifty. Dialed down like this, a hedgehog uses hardly any energy at all and can make it through to the next spring on its reserves.

Hedgehogs don't mind the cold—quite the opposite, in fact. The strategy I have just described works best when it is cold and frosty outside. It becomes dangerous only when winter temperatures rise above about 43 degrees Fahrenheit. Then the hedgehogs slowly rouse, and the deep sleep of hibernation turns into a lighter sleep during which considerably more energy is used but the animals are not awake enough to move. If such conditions persist for a long time, some of the little sleepers starve. It is only when temperatures reach about 53 degrees that the hedgehogs can really get moving and nibble on something—that is, if they are able to find anything because their prey are still hiding out in their winter quarters. With any luck, a number of early risers in this predicament will be found and fattened up in hedgehog rescue centers.

And what do hibernating hedgehogs dream of? During the really deep phase of sleep, metabolic rates are extremely low and there's hardly any dreaming going on, because the brain uses a lot of energy in the highly active state of dreaming. Therefore without metabolic activity, there are definitely no mental movies being played. But what about that light sleep above about 43 degrees Fahrenheit? If hedgehogs can dream then (after all, energy use rises), the pictures might be more like nightmares from which the hedgehog would like to wake but can't. Whatever is going on inside its head, the situation is life threatening and perhaps the animal realizes this in its drowsy state as it struggles in vain to wake up. Poor little thing. Climate change will, unfortunately, bring more of these warm winter interludes.

Things are somewhat better for squirrels, at least in the dream department. They don't hibernate, they just doze for two or three days at a time before they wake up again and feel hungry. Although their heart rate drops during these down times, so they use fewer calories, their body temperature remains high. This means they need regular servings of energy-rich foods such as acorns and beechnuts. If there are none or if the squirrels cannot find them, they starve. The strategy of red deer is much more like that of hedgehogs, because, surprisingly enough, they, too, can lower their peripheral body temperature. They do this multiple times during the course of a day, meaning that their periods of winter "hibernation" last only a few hours. Even such short down times allow them to restrict their use of valuable body fat. Despite low exterior temperatures, their metabolic rate is then up to 60 percent slower than in summer.[30]

Now another problem arises: red deer have to rev up their metabolic rate when they are digesting food, but going through the winter without eating anything is usually not an option, either. When a red deer eats in winter, it often extracts less energy from the food than it uses to digest it. And that is why, paradoxically, when hunters feed red deer in winter, the animals can end up starving in droves. This is what happened in 2013 in my home district of Ahrweiler, where there was an indignant outcry from hunters who wanted to continue feeding despite a local ban. Almost a hundred red deer died from starvation, a number of which would likely have survived if they hadn't been physically stressed after digesting the hay and sugar beets hunters had fed them. That is why, left to their own devices in winter, red deer mostly live off the body fat they accumulate in the fall.

At some point, I began to worry whether red deer constantly feel hungry in winter, which is a distressing thought.

To stand in cold snow with a rumbling stomach and super-cooled extremities is surely very unpleasant—at least it would be for a human being. It has now been proven that animals can turn off the sensation of hunger. Hunger is, after all, a signal from the unconscious that it's time to eat. And this feeling should only trigger the desire to eat when adding calories would be beneficial. Take hedgehogs, for instance. Even when they are hungry, they refuse smelly, rotten food. The unconscious part of their brain temporarily shuts down their hunger pangs and replaces them with a firm resolve not to eat any of the proffered sustenance. We don't know whether red deer experience an aversion to buds and dry grass or whether they simply feel full. But we do know that the animals don't feel hungry in winter despite their fast, because at the end of the day not eating makes fewer demands on their energy reserves.

The combination of lowering body temperature and metabolic rate that I have just described doesn't work equally well for all red deer. How well it works depends on the character of the individual deer and, equally importantly, on its rank and position in the herd. Winter is particularly dangerous for red deer that have strong personalities. As leaders in the herd, they have to be constantly alert. This means their heart rate is constantly elevated, and their energy use is correspondingly high. It's true that herd leaders have preferential access to good feeding grounds, but that is of little use to them. Meager winter offerings of dry grass and tree bark don't deliver sufficient calories to build up the fat reserves these deer are burning through at a higher rate than their lower-ranking herd mates, which spend the cold winter nights standing around calmly and dozing. The lower-ranking deer are eating less than their leaders, but they are also using far less energy. And so, at the end of winter, they have more reserves than their superiors. After watching deer in large natural enclosures, foresters in Vienna made the surprising discovery that being the leader of a herd reduces a deer's chance of survival, despite the fact that leaders always get to help themselves to food first. According to the researchers, in future it will be more important to consider the life histories and personalities of individual animals rather than the average for the species. After all, that is exactly how evolution works—using deviations from the norm.[31]

Warm-blooded and cold-blooded, then, are fluid categories that can merge into one another. What about feeling like you're freezing? This sensation signals to the body that its temperature is falling dangerously low and measures must be taken to counteract this. For us, it's lights out when our body temperature dips lower than 93 degrees Fahrenheit. Before that, we start to shiver and we try to get somewhere warmer. It's the same for our horses. On cold, windy winter days, our older horse, Zipy, is particularly prone to shivering, and she takes refuge in the pasture shelter. As the mare has less fat and muscle mass than her companion, and as her body is less well insulated despite her winter coat, that's sometimes not enough. On really cold days, we put a rug on her to warm her up until the shivering stops and she feels better. It's quite clear that feeling cold is as unpleasant for Zipy as it is for us.

But what is it like for insects? Their body temperature fluctuates along with the air temperature, and they don't have a mechanism to keep it at a certain level. In fall, insects burrow into the ground or under tree bark or into plant stems so they don't freeze solid. To ensure that any ice that forms in their cells doesn't cause the cells to burst, they store substances like glycerin, which inhibits the formation of larger and sharper ice crystals. What might that feel like? Do such creatures even experience cold? When I watch frogs and toads jumping into ice-cold ponds in late fall in order to doze in the mud at the bottom, I can't imagine that they feel freezing cold. We find cold water so unpleasant only because it draws off our body heat far more efficiently than air. But if your body temperature is the same temperature as the water in the pond, it can't feel too bad to jump in. And so the frogs and toads down there are probably not feeling like they're freezing after all.

But if insects, lizards, and snakes don't know what it's like to feel uncomfortably cold, does that also mean they don't know what it's like to feel pleasantly warm? I don't believe that. After all, in spring, these creatures love to seek out sunny spots. The more their small body heats up, the more swiftly they can move. Therefore, they perceive being warm as something positive, a condition that can be costly for some. Take slow worms, for example. (Slow "worms" are actually a kind of legless lizard.) Roads warm up particularly quickly in the sun. Asphalt stores heat and radiates it, even at night, so roads are a good place to fill up on warmth. Unless, that is, a car runs over the little sun worshipper, which, unfortunately, happens quite often. Drama aside, it is clear that, like us, cold-blooded animals must also have a feeling for temperature, although they probably sense it differently than we do.

12

CROWD INTELLIGENCE

SOCIAL INSECTS BELIEVE in division of labor. Early on, scientists coined the term "superorganism" to describe a collective in which each individual is part of a greater whole. In the woods where I live, red wood ants with their enormous ant hills are a typical example of this kind of arrangement. The largest ant hill I ever found measured 15 feet from side to side. There are usually multiple queens inside laying eggs to ensure the survival of the colony. The queens are cared for by up to a million workers, all female. The lowest on the social ladder are the winged males that fly out to mate with other queens and then die. The female workers, with a life span of up to six years, are unusually long lived for insects, but the queens, which live for up to twenty-five years, eclipse even this impressive achievement. Not in the literal sense of the word, I have to say, because ants need the warmth of the sun so they can work. That's why they live in bright, airy coniferous forests.

In Central Europe, red wood ants have spread beyond their natural range thanks to the proliferation of spruce and pine plantations. The fact that these ants are legally protected has less to do with their scarcity than with their reputation as "wood police." Supposedly, they help foresters rid forests of troublesome pests such as bark beetles and caterpillars, although the red-black ants are not the least interested in their appointed role. In addition to the aforementioned pests, they naturally also eat protected and extremely rare species. As ants, they have no concept of our human habit of categorizing creatures as either "pests" or "beneficials," but that doesn't make their colonies any less fascinating.

Their relatives, the bees, live in a similar fashion and have been particularly well researched. Bees are also born into a society with a strict division of labor. There's the queen, which develops from a normal, fertilized larva. Whereas other baby bees are fed a mixture of nectar and pollen, larvae destined to hatch into royalty receive a special food called royal jelly. It's produced in the hypopharyngeal glands of worker bees, and while normal larvae develop into mature bees in less than twenty-one days, the turbodiet of jelly produces a new queen after just sixteen days. A queen flies only once in her life: her nuptial flight. And it is during this flight that she mates with drones (a.k.a. male bees). After her return to the hive, for the rest of her life (about four or five years), she lays up to two thousand eggs a day, interrupted only by short winter breaks.

For their part, the female worker bees spend every moment of their short lives hard at work. In the first days after they hatch, their job is to feed the larvae. After ten days, they are also responsible for storing nectar and converting it to honey. Once three weeks have passed, they are finally allowed out into meadows and pastures to collect nectar for another three weeks. Then they are worn out and die. Only the overwintering bees that wait for the next spring huddled together in a tight cluster around the queen get to live a little longer. The only task for the drones is to inseminate queens, and because that happens only once and only a few of them get this opportunity, most of the time they just hang around with nothing much to do.

Everything, then, is strictly preprogrammed down to the smallest processes. Inside the hive, the bees dance to communicate information about nectar sources and how far away they are. They turn nectar into honey by adding glandular secretions and drying the mixture on their tiny tongues. They sweat out wax and form it into artful combs. Scientists recognize the bees' accomplishments, but because, in their opinion, small insect brains are unable to rise to intellectual heights, the individual bees are considered to be components of a superorganism and their cognitive accomplishment is called crowd intelligence. In such an organism, all the animals are like cells working together in a much larger body. Whereas the individual animals are considered to be quite stupid, the interaction of the different processes, as well as the ability of the whole to react to stimuli in the world around it, is recognized as intelligent. This perspective does not recognize bees as individuals; instead, each bee is reduced to a building block, a piece of a larger puzzle. Consequently, in the language of German beekeepers of old, the colony of bees was called "the bee," referring to the population of bees as though it were a single entity.

Whatever we think, it's all the same to the bees. And since I've been keeping bees, I now know that this point of view is incorrect, because there's a lot more going on inside their little heads. For example, bees can definitely remember people. They will attack people who have annoyed them in the past, and allow people who have left them in peace to venture much closer. Professor Randolf Menzel at the Free University of Berlin has discovered other amazing things. Young bees leaving the hive for the first time use the sun as a kind of compass. With the sun as their guide, they develop an internal map of the landscape around their home and use it to record their flight paths. In other words, they have an idea about what things look like around them. In this, they orient themselves much the same way we do, for people also create mental maps.[32]

But that's not all. The waggle dance returning worker bees perform for other bees indicates the abundance of, the direction of, and the distance to a source of nectar, such as a field of rapeseed in full bloom. In his experiment, Randolf Menzel and his colleagues saw to it that the nectar source was then removed. When the frustrated bees returned, they got new coordinates by watching the dance of workers who had spotted blooms elsewhere. But the researchers then removed this second source, as well, which meant more frustrated returnees. After this, Menzel noticed something completely unexpected. Some bees tried the first location a second time and, when they discovered there was still nothing there, they flew directly to the second location. But how did they manage that? All the waggle dance had communicated was distance and direction from the hive. The only explanation is that the little creatures actively processed the information about the second location so they could find it from the first. You could say they remembered, reflected, and then rerouted. Crowd intelligence could be of very little use to them here, and they must have relied on their own little brains to make this calculation. And there's more. Insofar as they plan for the future, reflect on things they have not yet seen, and are aware of their bodies in relation to these thoughts, they are self-aware. "The bee knows who it is," says Randolf Menzel. And for that, it has no need of a crowd.[33]

13

HIDDEN AGENDAS

I F BEES KNOW who they are and plan for the future, what about birds and mammals? Whenever I'm animal watching, I wonder whether the individuals I'm observing are consciously aware of what they're doing. That's very difficult for a layperson to determine—and that is what I am, despite my engagement with the topic—but as I explore the subject, I don't want to rely on studies alone; I also want to experience firsthand what animals are thinking. That might sound like asking for too much, because it's difficult enough to work out what other people are thinking just by watching them, but during a conversation at the breakfast table, my children suggested I had already experienced something along these lines, albeit only briefly.

I was telling them about the crow that waited for us every morning in the horse pasture. The black bird was always hanging around out there with a few other crows, and its territory must have been close by. Unfortunately, it's still legal to hunt and shoot crows where I live, and so these intelligent birds are very shy around people and normally keep a safe distance, usually about 100 yards. Over time, the crows in the pasture have become used to us, and they consider a distance of about 100 feet to be sufficient. That is, all except one, which is gradually getting tamer. On a good day, it allows us to approach within 15 feet, and every time, we are touched by its trust. We talk to it, and it always gets a bit of grain, which we put on the hitching rail by the pasture gate. Aha—food. No, the crow doesn't approach because it's curious or simply wants to be with us. It's well aware that when we appear, so does a meal. Even so, we enjoy seeing the crow on our daily rounds. We don't set the emotional bar very high, and that's just fine with us.

Because of this, on the morning in question, I observed something that I found merely amusing at first. It was December. The pasture was soaked after weeks of rain and every step I took in my heavy rubber boots sent up a spray of mud. It's hardly ever any fun going out to feed the animals in weather like that, especially not when a side wind is blowing a persistent drizzle into your face. Oh well. The horses were waiting for their morning ration of grain, and it's a well-known fact that moving around in fresh air always makes you feel better. To prevent the younger mare from immediately polishing off the older mare's portion, I had to wait, ready to run interference if Bridgi started helping herself to Zipy's food. Usually my presence was enough to ensure that the younger animal behaved herself, and in the minutes it took the horses to eat their breakfast, I had time to watch the landscape—or rather, the crow.

That morning, the black bird flew over from the nearby wood, having spied me a while ago in my green-and-orange jacket with the white feed bucket in my hand. However, instead of flying to its usual observation spot, a post near the hitching rail, it landed in the pasture only about 60 feet away. I suddenly noticed it was carrying something in its beak: an acorn. The crow then tried to hide its tasty treat. It poked a hole in the ground, stuffed the acorn into it, and pulled a tuft of grass over the hole. I was marveling at the perfect camouflage when the crow turned in my direction. Had it noticed that I was watching? Immediately, it retrieved the acorn from its hiding place and began to poke another hole in the ground. Just the one? No, many holes, and at each one, the crow pretended to stuff the acorn inside. When it came to the last hole, the nut disappeared and the bird was satisfied. After all, it had gone to a great deal of trouble to deceive me and prevent me from eating its favorite food. Only then did the crow fly over and land on the rail to eat the small offering of grain I had left for it.

When I related this little story back at home over breakfast, my children pointed out that it was a fine example of forward planning. Then the penny dropped. The whole time I had been amused at how the bird was hiding its food from me, and that alone was highly intelligent behavior. After all, the bird had to consider what I might have seen and how, when it was being observed, it could hide the acorn in such a way that

I would be fooled. But right there in front of me, the crow had considered something else, as well.

Even a crow's stomach can hold only so much, and clearly an acorn meal would have filled it up. Of course, the crow could still have flown over to the offering of grain, but as it would already be full, its best option then would be to hide the grain. However, hiding individual kernels of grain is very labor intensive, and so even though it was hungry, the bird first stashed the large acorn safely and then flew over to the rail to fill its stomach at leisure. Finally, it joined its companions in the neighboring pasture, and I'm sure it retrieved the acorn later. That was perfect time management to optimally exploit the offer of food, and to do that the crow must have given thoughtful consideration to the future. And, for me, it was a great incentive to look more closely the next time I observed animals and, above all, to think more carefully about what I was seeing. And who knows, perhaps you, too, have stumbled across such experiences, and now you can puzzle them out with the benefit of hindsight.

14

SIMPLE SUMS

I N MY BOOK *The Hidden Life of Trees*, I reported that trees can count. In spring, they make a note of the warm days above 68 degrees Fahrenheit, and they leaf out only when a certain threshold has been exceeded. If these large plants can count, it makes sense to assume that animals have this ability as well. People have certainly wanted them to be able to count for a long time, and there have been many reports of amazing animals. Take Clever Hans, for example. This stallion could spell out words, read, and count—or at least that is what his owner, Wilhelm von Osten, claimed when he made the horse a public attraction in Berlin early in the twentieth century. An investigative committee confirmed the horse's abilities without being able to explain them. Eventually, the puzzle was solved. Clever Hans was reacting to his owner's almost imperceptible facial expressions. The moment that von

Osten disappeared from the horse's sight, the horse's abilities disappeared as well.[34] By the end of the twentieth century there was an increasing amount of hard data that confirmed that many animals can perform some rudimentary calculations. Mostly the data had to do with food and estimating larger or smaller quantities. Personally, I don't think it's particularly interesting to acknowledge that animals can tell the difference. After all, when faced with the choice of more or less, deciding for more is part of what drives evolution, isn't it? What is much more interesting is whether real counting is possible.

Perhaps we can get closer to an answer by considering our goats. This time it wasn't me but my son who discovered what might be going on in the minds of Bärli, Flocke, and Vito. While we were on vacation, Tobias took over responsibility for our little Noah's ark. Usually the goats get a small ration of grain at noon, and this is the highlight of their day. When it's time for their snack and we're in the pasture, the eager goats come running. In contrast, in the morning and evening, when we are "only" feeding the horses, which are right there next to them in the pasture, the goats barely acknowledge our presence. Tobias changed the feeding times to suit his schedule, and every day was different. Sometimes the goats weren't fed until early evening and the horses' last feed was even later. When Tobias made his second visit to the pasture in the early evening, Bärli and her little family ran to him bleating, noisily demanding their portion of grain. It was, after all, my son's second appearance of the day, even though it did not coincide with the time of day when the goats were expecting a tasty treat. Is that evidence that goats can count? They always love grain, but they were now demanding it at what was for them an unusual time. Did they know that this was Tobias's second visit to the pasture and so it was their turn to be fed? If it was nothing more than gluttony, then they would have begged, as so many domestic animals do, and demanded food every time a member of the family appeared. But they did so only on one out of three daily visits—the middle one.

Apart from our goats, what does this evidence of intelligence look like in other creatures? It's pretty much old news by now that ravens are in the same league as apes, so let's take a look at pigeons instead. These birds have become a plague in our cities, and I must admit that it's not very nice to have a dollop of bird poop land on your new jacket while you're standing on the sidewalk, which is what happened to me only recently. Despite the mess, the birds don't deserve the popular pejorative "rats with wings," and it is thanks to their intelligence that they can establish themselves so persistently in our urban pedestrian zones. Professor Onur Güntürkün of the University of the Ruhr-Bochum has amazing tales to tell. His colleagues trained pigeons to recognize cards with abstract patterns. After training, the birds could differentiate between a staggering 725 different images. They were divided into "good" and "bad" cards that were presented in pairs. If a pigeon pecked at the good card, it got food. If it sank its beak into the bad one, not only did it not get any food, but it also found itself in the dark (which pigeons can't stand). To pass the test, all the birds had to do was to memorize just the good cards; however, the scientists ascertained by means of a control that the birds were not taking the easy way out but actually memorizing all the pictures, both good and bad.[35]

Our dog Maxi offered a completely different example of the ability to count, and it had to do with her ability to tell time. She had the habit of sleeping soundly through the night until just before 6:30 a.m., when she would begin to whine softly, asking me to take her walkies. Why 6:30 a.m.? This was the time our alarm always went off and the whole family got up to eat breakfast and then leave for school or work. It seems we could have managed fine without an alarm of our own as Maxi had a well-functioning internal clock, albeit one that clearly ran five minutes fast. But it was different on weekends. The alarm was switched off and we could all sleep in. All of us. For there wasn't a peep out of Maxi on Saturdays and Sundays. In fact, she slept in longer than we did—great evidence that dogs can count. Now, you could argue that the dog had observed our behavior and registered the fact that we sleep longer on weekends. But you can rule out that line of reasoning because during the week she always woke us before our alarm went off while we were all still asleep. In the same situation on the weekend, however, she didn't do that. We never did figure out why she stayed in her basket then and slept even longer than the rest of us.

15

JUST FOR FUN

ARE ANIMALS CAPABLE of having fun? That is to say, are they capable of doing things that have no particular purpose other than to bring them pleasure and happiness? I think that's an important question, because the answer helps us decide whether animals experience positive feelings only when they perform tasks that promote the survival of the species (such as pleasure during sex, which creates offspring). If that were the case, then pleasure and happiness would be byproducts of purely instinctive programming that ensures certain behaviors are engaged in and rewarded. In contrast, just by remembering happy experiences, we can relive the emotions that went with them and enjoy them over and over again. Free-time fun belongs here, such as a vacation by the sea or winter sports in the mountains. Might this be the crowning glory that sets us apart from animals? But then those tobogganing crows come to mind. An internet video shows one of these birds sliding down the roof of a house. The bird has a lid from a plastic container. It carries it up to the highest part of the roof, places it on the slope, and then jumps onto it to slide down. No sooner does the bird reach the bottom than it goes back up for its next ride.[36] The point? Apparently none. The fun factor? Probably the same as when we jump onto the wooden or plastic object of our choice and careen down a hill.

Why would crows expend energy on a pointless activity? After all, tough evolutionary competition calls for the elimination of all non-beneficial activities and ejects from the race any animal that is not sufficiently rigorous in this respect. And yet it's been a long time since we've paid any attention to this seemingly absolute rule. At least in wealthier countries, we have energy to spare, and we can afford to use it to enjoy ourselves. Why should it be any different for an intelligent bird that has set aside sufficient food for the winter and can devote some of these calories to fun and games? Clearly crows, too, can convert surplus resources into mindless fun and conjure up happy feelings whenever they want.

And what about dogs and cats? Anyone who lives with these animals can tell stories about how they love to play. Our dog Maxi liked to play tag with me around our forest lodge. Because she knew she could run much faster than I could, she always gave me a chance to catch her so the game didn't get boring. She'd run big circles around me, every once in a while dashing toward me. Then, just before I caught her, she'd sidestep and I'd miss. You could tell just by looking at her how this game delighted her. I really enjoy looking back to that time, and yet I'd rather find other examples as evidence of completely pointless play—meaning pointless in a positive way—because Maxi probably used this game to strengthen our relationship. And it's true that any playful activity within a group can act as social glue and therefore serve an evolutionary purpose. Energy invested in cohesion promotes groups that are particularly resistant to external threats.

So let's take another look at crows. There are lots of reports of crows that tease dogs. They stalk them from behind and nip them on the tail. Of course, the dog spins around too slowly to catch the bird, which soon starts the game all over again. This is not a case of creating social cohesion, and it's not a case of the bird honing some survival skill, either. After all, escaping from spinning dogs is not a necessary part of its behavioral repertoire. No, what's going on here seems to be something completely different. The crow can clearly put itself in the dog's place and realizes that the dog will always be too slow and will therefore get annoyed. And that's what makes it so much fun to tease it over and over again, gleefully anticipating its reaction. Lots of crows enjoy doing this, as any number of internet videos attest.

16

DESIRE

EX IS NOT an automatic process for animals, although you'd be forgiven if after reading scientific papers on "mating," you concluded that it is an act completely devoid of emotion. Hormones feature, and they trigger instinctive behaviors that animals are incapable of resisting. But is it any different for people? I just have to think back to the couple I came across years ago in the woods. All I wanted to do was to check out whose car was parked in the undergrowth, when two beet-red faces appeared from behind the hood. I knew both the man and the woman. They came from neighboring villages and at the time were married to different partners (and they still are). They quickly adjusted their clothing, climbed into their cars without saying a word, and disappeared. Clearly, they didn't want to risk their marriages and therefore had chosen a supposedly secluded spot for sex.

Even though there was naturally still a risk of dire consequences for their personal lives, both of them had succumbed to desire. I think this is a good example of how we, too, are at the mercy of our instincts.

The trigger for such behavior is a cocktail of hormones that gives rise to feelings of exquisite satisfaction and joy. Why is that even necessary? If living beings are supposed to mate, they could mate as instinctively as they breathe. After all, our bodies don't release supplements of drug-like substances to trigger every breath. Mating is special because during copulation all species abandon themselves to a state of helplessness. Some animals are sadomasochistic, such as snails, which ram chalky darts into each other's body to stimulate their partner. Male peacocks and grouse attract their hen's attention with a fan of erect tail feathers before jumping on her. Insect couples hitch aerial piggyback rides, while male toads in the raptures of love clamp down on their female underwater. Sometimes multiple males pile up on top of one another and don't let go, holding the female down for so long that she drowns.

Goats behave in many ways like deer, and every year in late summer, we observe a somewhat more elaborate ritual. At that time of year, our billy goat, Vito, turns into a monster of stink. To please the ladies, he perfumes his face and forelegs with a very special fragrance: his own urine. And he doesn't just spray the yellow liquid onto his skin; he also sprays it into his mouth. What would make us gag obviously doesn't fail to have the desired effect on the female goats. They rub their heads over his coat to absorb the smell. This clearly stimulates hormone production in all involved, and the fires of desire are lit. The billy goat keeps testing with his nose to see if one of the does is ready to let him in. He does this by driving her around the pasture while bleating with his tongue hanging out, which actually looks a bit ridiculous. If the lady of his dreams stands her ground and squats to pee, he shoves his nose into the stream. Then he snorts and curls back his upper lip to check her hormone levels to see if this will be his lucky day. After many days of testing, the does finally allow Vito a few seconds of bliss.

Let's get back to the question of why there has to be any sort of hormone-fueled emotional reward. The backstory is that mating is dangerous. The prelude to mating, when males often draw attention to themselves, doesn't just attract females. No, indeed. Hungry predators are also grateful for an eye- or ear-catching sign pointing the way to a tasty meal. And quite a few males of many different species do, in fact, step off their woodland stage directly into the stomach of a bird or a fox. The actual act of mating is even more dangerous. While it's going on, both participants are pressed tightly together for a few seconds and sometimes even for many minutes and are hardly in a position to flee from an attack.

We don't know if animals make the connection between mating and offspring, yet what other reason could there be for them take this risk? Only one comes to mind: the strong, addictive feeling of orgasm that causes them to throw caution to the wind and abandon themselves to pleasure. And therefore I believe, without a doubt, that animals experience intense emotions when they have sex. And there is another strong indicator that this is so. Quite a few animals have been observed pleasuring themselves. Deer, horses, wild cats, or brown bears—all have been seen either laying a hand (that is to say a paw) on themselves or taking advantage of natural aids such as tree trunks. Unfortunately, there are not many reports, let alone any research, on the subject, perhaps because masturbation is considered taboo.

17

TILL DEATH DO US PART

DOES IT MAKE sense to think of animal pairings as marriage? According to the dictionary definition, marriage is a legally recognized lifelong partnership between two individuals. *Wikipedia* states that marriage is "a socially or ritually recognized union or legal contract between spouses."[37] There is no such thing as legal recognition for animals, but there is such a thing as a close lifelong partnership. The raven is a particularly moving example. It is the largest songbird in the world, and by the middle of the twentieth century, it was almost extirpated in Central Europe. It was alleged that ravens killed grazing animals as large as cows. Today we know that is nonsense. Ravens are the vultures of the north, and they seek out dead or at most dying animals. They were ruthlessly hunted and killed using both weapons and poison.

Historically, campaigns to get rid of undesirable animals have met with mixed success. In the twentieth century, people wanted to get rid of the red fox because it can carry rabies. Foxes were shot on sight (and they still are), their dens were dug up, and any young found in them were destroyed. Poison gas pumped into their underground accommodations was a convenient way to kill them. But despite these measures, the red fox has survived, because it is very adaptable and reproduces prolifically. And, most importantly, red foxes change their sexual partner. In contrast, ravens are faithful souls and stick with their mate for life. In this respect, we can indeed talk of a true animal marriage. During the extermination campaign, the birds' fidelity was their undoing. If one bird in a pair was shot or poisoned, the other often didn't seek out a new partner, but from then on flew its circuits in the sky alone. The large number of single birds contributed no young, which hastened destruction of the species. Today, ravens are strictly protected in Central Europe and once again populate their ancestral range. I still remember a trip Miriam and I made to Sweden with our children. As we paddled our canoes over a deserted lake we often heard ravens call, and I was entranced. I was so excited when a few years ago I heard these birds for the first time in the forest I manage in Hümmel. Since then, I think of these animals as symbols for how Nature can heal from our crimes against it, and how destruction of the environment need not be a road of no return.

Monogamous animals are common, especially among birds. There are several species that behave like ravens, although none quite as rigorously. They remain true to the same partner, at least within a single breeding season. There's the white stork, for instance. Although what the stork remains true to is actually its nest from one season to the next, and often the only reason partners meet up again is because both head back to their old nest every spring. Sometimes, however, things can go awry, as reported by an employee at the Heidelberg Zoo. That spring, a stork built a nest up high with a new female—his former partner having obviously gone astray during migration—but then, in the middle of this cozy togetherness, his former partner belatedly appeared. Now the male stork found himself in a quandary. In order to be fair to both females, he built a second nest and then ran himself ragged providing for both families.[38]

But why aren't all species of birds so loyal? And what does being loyal really mean? Just because chickadees and other species don't mate for life doesn't mean they are disloyal partners. The reason for a bond that lasts for just one season lies in the average life span of the birds. Whereas ravens even in the dangers of the wild live for more than twenty years, for other species, mostly the smaller birds, it's all over after less than five. If you mate for life and the likelihood your partner will go missing is very high, then there would soon be mostly singles flying around. And because that would be exceedingly bad news for the survival of the species, for birds with short life expectancies there's a new roll of the dice every spring to see who partners up with whom. That's when the birds can see who survived the winter and migration. Great tits and robin redbreasts waste no time grieving over a mate from the previous year that never returns.

And what about mammals? There are only a few that hook up like ravens do. Beavers are one such example. They find a life partner and remain with the same animal for up to twenty years. Their children don't move out immediately, and the young continue living with their parents in comfortable earth lodges close to the water until they are two years old. Most other species are somewhat challenged in the relationship department, at least when it comes to the opposite sex. With red deer, it's a case of "to the strongest go the spoils." Once a mighty stag has routed his rivals, he satisfies himself with the harem of does until he, in turn, is driven off by an even stronger male. It's clearly all the same to the does. They allow themselves to be mounted by any young buck that tries his luck when the alpha male is not paying attention. Raising fawns is purely women's work anyway, because by that time, the fathers are already off wandering through the woods in male-only groups.

18

WHAT'S IN A NAME?

IT SEEMS OBVIOUS to us that we can address each other by name when we communicate. In large societies, each person we want to single out has a personal name we can use when we want to get their attention. Whether we're using email, an online app, or the phone, or having a face-to-face conversation, without this direct form of address we'd be lost. We're conscious of how important this is when we meet someone we've been introduced to before and we forget their name. Is naming a particularly human habit or is there something similar in the animal realm? After all, all social creatures face the same issue.

In mammals, there is a simple form of naming between mother and child. The mother utters a call using her normal voice. The child recognizes the call and responds with a clear vocalization of its own. But is this really an exchange of names or is this merely voice recognition? The argument in favor of the latter interpretation is that these special mother-child "names" seem to fade with time. Once the young animals are grown up and weaned, their mother no longer responds to them. What meaning does a name you've given yourself have when no one reacts to it? Does a temporarily meaningful call deserve to be called a name?

Even if we disallow such calls, science has discovered some instances of genuine personal naming in the animal kingdom. It's no coincidence that this naming happens, once again, with ravens. Their close societal bonds provide an ideal background for answering questions about naming, because ravens cultivate lifelong relationships, not only between parents and children, but also among friends. Naming calls are the way to go if you want to communicate over long distances and identify which individual is doing the talking. These inky black birds can master more than eighty different calls, a raven vocabulary if you will. Amongst these is a personal identification call a raven uses to announce its presence to other ravens. But is this call really a name? That would only be true in the way we use the word if other ravens also "addressed" the speaker using its personal identification call—and that's exactly what ravens do.[39] They remember the names of other ravens for years, even if they've had no contact. If an acquaintance appears in the sky and calls his or her name from afar, there are two possible responses. If the returning raven is a former friend, the other ravens answer in high-pitched, friendly voices. However, if the raven is unpopular, the greeting is low-pitched and brusque. Similar observations have been made for human greetings.[40]

It's quite difficult to discover the names animals use for each other. It's much easier if we call them using a name we've chosen for them and see if they react. People who own a single pet now come to the next hurdle. How do we know that, for instance, our dog Maxi doesn't hear her name and interpret it to mean simply "Hi" or "Come here"? It would be easier to tell if you owned more than one dog, but here I would like to return to the intelligent pig. Researchers have studied pigs for precisely this attribute. The reason for the research was the widespread pushing and shoving in modern pens. Farmers used to pour feed into a long trough so all the pigs could eat at the same time. Today, feed is distributed automatically with the help of computers that calculate the amount for each individual pig, and because the equipment is expensive, there's not enough money for many machines and so not all the pigs in one pen can eat at the same time. They have to wait their turn, and pigs are as impatient with rumbling stomachs as we are. They jostle each other in line, sometimes even injuring each other.

To help bring some civility back into the process, researchers from the Friedrich-Loeffler-Institut (the German Federal Research Institute for Animal Health), specifically from the institute's pig team, tried to teach manners to pigs at an experimental farm in Mecklenhorst in Lower Saxony. Here, using small school "classes" of eight to ten animals, they trained yearlings to respond to individual names. The young porkers were particularly good at catching on to three-syllable female names. After one week of training, the pigs went back into a pen in a larger group. And now meal times got really interesting. Each pig was called up by name when it was its turn. And, as it turns out, it worked. As soon as "Brunnhilde" rang out from the loudspeaker, the only pig that sprang up and raced to the trough was the one that had been called, while all the others continued doing whatever it was they were doing, which for most of them was simply snoozing. The measured heart rate of the other pigs did not increase, and the only pig that registered a higher pulse was the one that had been summoned. The new system had a success rate of 90 percent, and is one way to bring peace and order to the pens.[41]

But does this heartwarming discovery have a wider significance? The ability to associate oneself with a particular name presupposes self-awareness. And that is one step above consciousness. For whereas the latter merely implies thought processes, with self-awareness we're talking about recognizing who you are, a sense of self. To test whether animals possess this ability, science came up with the mirror test: animals that recognize that the reflection in the mirror is not another animal but their own image are supposedly self-aware. The inventor of this experiment was one Gordon Gallup, who painted a colored spot on the forehead of anesthetized chimpanzees. Then he placed a mirror in front of the drugged animals and waited to see what happened when they woke up. No sooner did the apes glimpse their reflections through eyes still blurry with sleep than they began to try to rub the color off. Clearly they understood immediately that they were the ones looking out from the gleaming glass. Since then, this test has been considered proof that animals that pass it possess self-awareness. (Incidentally, children do not pass this test until they are about eighteen months old.) Primates, dolphins, and elephants have all passed the test since it was first introduced and, accordingly, have risen in the eyes of researchers.

People were surprised when crows also recognized their own reflection, as did magpies and ravens. Thanks to corvids' intelligence, people have started referring to these birds as "feathered apes."[42] Not much more came to light in connection with this discovery for a long time, until suddenly pigs began popping up in scientific papers. Pigs? They, too, passed the test, although unfortunately they didn't score a pet name ("factory farm apes" comes to mind) for if they had, how could anyone have continued to treat pigs as inhumanely as they are treated today? People don't even credit these intelligent animals with feeling pain, as proven by the fact that, in Germany, it is legal to castrate piglets only a few days old without anesthetic, and it will continue to be legal until 2019, because it's quicker and cheaper that way.

But back to the mirror. Pigs know how to use mirrors for more than just self-contemplation. Donald M. Broom and his team at the University of Cambridge placed food behind a barrier. Then the pigs were positioned so that they could only see the food using a mirror placed in front of them. Seven out of eight pigs took just a few seconds to realize that they had to turn around and go behind the barrier to reach the tasty treats. To do this, they not only had to recognize themselves in the mirror but also had to consider spatial relationships in their surroundings and their own place in them.[43]

Despite these results, we should be careful not to set too much store by the results of the mirror test, particularly when it comes to animals that do not pass it. First off, when dogs anointed this way look at their reflection and do not react, that may not mean anything. How are we to know whether the spot on their face bothers them in the slightest? And even if it does, perhaps they don't know how the mirror works. Perhaps they think it's a colorful picture or, at most, a video like the ones we watch on a television.

Let's return to naming, and now those Canadian squirrels take center stage again. When investigating cases of adoption, researchers discovered that these arboreal imps adopted only related young. But how did they know which ones were their nieces, nephews, or grandkids? Researchers at McGill University suspect that adults' vocalizations have an important role to play here. Every squirrel has a distinctive call, and the solitary creatures use these calls to recognize one another. After all, they don't see each other very often because their territories rarely intersect, so the only way they can communicate is by sound. What is even more astounding is that some animals come looking when they stop hearing their relatives' calls. This means they have to leave their own territory and enter an unfamiliar one. Does that make them nervous? We can only speculate about that, but we do know that when they come across orphaned young during their forays into alien territory, they take the helpless young into their care.[44]

Science is only just beginning to explore this subject, like so many others. As I've just explained, naming is an advanced form of communication that many animals have mastered. Even fish, which we think of as silent creatures, are in on this skill, but the only thing we know right now is that they use the sounds they make to find a partner or to defend their territory.[45]

19

GRIEF

RED DEER ARE social animals. They congregate in large herds and enjoy being in a crowd. That being said, there is a definite difference between the sexes. After they turn two, male red deer become restless and wander far afield. They get together with others of their sex, but they stay only loosely connected. As they mature, they become solitary and prefer to be on their own, occasionally tolerating the presence of a younger male in their vicinity. German hunters call these hangers-on "adjutants."

A female red deer is much more constant. Her herd is a tightly knit community led by a particularly experienced alpha doe. The lead doe passes along to the younger females established behaviors passed down to her by her predecessors, including the use of decades-old travel corridors between territories. These are paths the herds use to reach lush summer meadows or protected winter quarters. When danger threatens, the frightened females look to their leader. She's the first to know what to do, because she can remember similar situations and potential predators. These dangers don't just come from animals. For example, I have often observed a herd of deer leaving an area at the start of a drive hunt. A few blasts on the trusty hunting horn signal that it's time to get going. The call to action warms the hunters' hearts before they set off—and warns the lead deer that it is time to depart. This proves deer can remember the specific sequence of notes even though more than a year has passed since the last hunting season.

Besides age and experience, an alpha doe needs one more attribute to prove she deserves her position: offspring. Offspring are essential to show that the doe can take responsibility not only for herself but also for others. Some wildlife researchers attribute the fact that the other deer in the herd follow her to happenstance. The way they see it, because the deer feel comfortable only when they are in a group and as the older doe is leading her fawn anyway, they join her more or less randomly merely because two animals happen to be moving off in the same direction. I, however, am convinced that the members of the herd are well aware that a particularly experienced doe is taking the lead. She decides first, and the others are happy to defer to her decision. Researchers object to this, saying older animals are more vigilant and that's why she's the first to react when it's time to flee. It's hardly surprising that the others choose to follow her, to be on the safe side. What's going on is simply passive leadership and not active guidance.[46] I don't believe that, either. It's true that does don't fight for dominance in the herd, instead resolving issues of hierarchy quietly in ways we don't yet recognize. However, if leadership were merely a matter of chance, the does would follow one female one day and another the next—perhaps even an especially fearful doe, one that is not only young and inexperienced but also especially nervous, making her particularly prone to flight. But true leadership is characterized by something completely different: the ability to not get worked up unnecessarily, because an animal that panics too often has less time to eat and therefore less energy to ensure survival. It is because the other does recognize the experience that comes with age that they all quietly agree amongst themselves that the lead doe is the one they will follow.

There are times, however, when disaster strikes the alpha doe, for example when her fawn dies. In earlier times, the cause of death was usually disease or a hungry wolf. These days, it is often a blast from a hunter's gun. For deer, the same process is set in motion as for us. First, unbelievable confusion reigns, and then grief sets in. Grief? Are deer even capable of experiencing an emotion like that? Not only can they, but they must. Grief helps them to say goodbye. The bond between doe and fawn is so intense that it cannot be severed from one moment to the next. The doe must slowly accept that her child is dead and that she must distance herself from the tiny corpse. Over and over again, she returns to the spot where her child died and calls for her fawn, even if the hunter has carried it away. A grieving alpha doe endangers her kin by remaining close to the scene of death, which also means close to danger. What the herd needs to do is to move to a safer place, but departure is delayed until the bond between mother and fawn finally dissolves. There is no question that under such circumstances, it is time for a change in leadership, and this happens without any struggle for dominance. Another similarly experienced doe simply steps forward and takes over leadership of the group.

If it's the other way around and the alpha doe dies, leaving her fawn behind, the remaining does show no mercy. There is no question of an adoption. Quite the opposite. The orphaned youngster is frequently banished from the herd, perhaps because the other does don't like dynasties. Left to fend for itself, the fawn has little chance of surviving and usually doesn't make it through the next winter.

20

SHAME AND REGRET

NEVER WANTED TO have horses. I thought they were too big and too dangerous, and I was never interested in riding. That is until the day we bought two of them. My wife, Miriam, had long dreamed of living with horses, and there was enough pasture for rent close to our forest lodge. So when a horse owner who lived not too far away wanted to sell his animals, the ideal moment seemed to have arrived. Zipy, a quarter horse mare, was six and broken for riding. Her companion, a four-year-old Appaloosa mare called Bridgi, was considered unrideable because she had a diagnosed problem with her back. That was perfect. It had to be the two of them because herd animals should not be kept alone, and the fact that only one of them could be ridden was just fine with me, because I was out of the picture as far as riding was concerned anyway.

But then everything changed. Our vet examined both animals and decided there was nothing wrong with Bridgi after all. What was to stop her from being ridden as well? Nothing. And so Bridgi and I began to learn together with instruction from a riding coach. As I rode her, but even more as I cared for her daily, I developed a very close relationship with the mare until my fear disappeared completely. Moreover, I learned how sensitive horses are and how they react immediately to the slightest cues. If my wife or I were not on task or were on edge, they ignored our commands or shoved inconsiderately when we were feeding them. It was the same when we were riding them. The horses knew just by how tense our bodies were whether our cues (for instance, a small adjustment of weight in the direction we wanted them to go) were to be taken seriously or not. Conversely, in time we also learned how to pick up on very small cues from Zipy and Bridgi. And while we were handling the horses, we discovered the depths of their emotions.

For example, our horses have a highly developed sense of fairness, which shows itself in all kinds of situations. It's particularly marked and most easily observed at feeding time. Zipy, at the age of twenty-three, no longer digests pasture grass very well and would slowly waste away if we didn't step in to improve things. So every day at noon, she gets a large scoop of a concentrated grain-based feed. If Bridgi, three years her junior, has to watch, she becomes petulant. She dances around and lays her ears back (a guarded threat). In short, she sulks. So we throw her a handful of feed, which we spread out in a long line on the grass. Because she has to separate the feed from the blades of grass, she spends just as long eating as her older companion with her larger portion in the manger. And that sets the world to rights as far as Bridgi's concerned.

We noticed the same thing with training. The horses clearly enjoy being ridden in the small riding ring. And not just because of the exercise. They have plenty of that because they spend the whole year out in a large pasture. No, what the horses really enjoy is the attention we pay them when we practice different moves, and the praise and pats we give them when something works out well.

As we spent time with our horses, we noticed the stirring of another emotion: they can feel ashamed or embarrassed just like we can. Despite her twenty years, the lower-ranking Bridgi sometimes behaves like a silly, empty-headed young thing. When she's in one of those moods, she won't come directly to us when we call but prefers to take another turn around the pasture at a gallop, or she tries to grab something before we've given the command: "Time to eat!" Then we reprimand her, which we do by making her wait just a little bit longer for her food until she starts behaving herself again. Normally, she takes the rebuke in stride, but if the older Zipy is watching, she turns her head away sheepishly and starts to yawn. You can tell just by looking at her how embarrassed she is. To put it another way, Bridgi is ashamed of her girlish behavior.

If you think of similar scenarios involving people, you'll notice this happens when another person is involved, and it is their presence that makes things awkward. Clearly it's the same for horses, and I believe that many social animals experience similar feelings. The underlying reasons for these feelings have not been researched in animals, but they have been in people, and they give us an idea why shame and embarrassment exist at all: the person in question has transgressed social rules. They turn red and lower their gaze to signal submission. The other members of the group see their misery and, as a rule, feel sympathy for them, which means that the person who made the misstep is usually forgiven. Ultimately, shame and embarrassment function as a kind of act of contrition—they are mechanisms for asking for forgiveness. It is still mostly assumed animals cannot feel ashamed or embarrassed, because to do so an individual must be in a position to think about their own behavior and the effect it has on others.[47] I am unfortunately unaware of any current research on this topic, but there is a related emotion that I can report on, and that is regret.

How often in life do we regret making a bad decision? Regret is an emotion that generally protects us from making the same mistake twice. That makes perfect sense, because it stops us wasting energy by repeating dangerous or pointless behavior over and over again. And if it makes that much sense, then it also makes sense to look for an emotion like that in the animal world. Researchers at the University of Minnesota in Minneapolis observed rats with this in mind. They built a special "restaurant row" for rats—a ring with four spokes leading to four different feeding zones. When a rat came to the entrance of one of the spokes, a sound indicated how long the wait for food would be: the higher the sound, the longer the wait. And now the rodents began to act like people. Some lost patience and went on to the next spoke in the hope they would be served more quickly. Sometimes, however, the sound was even higher there, meaning the wait time would be even longer. Now the animals looked wistfully back in the direction of the spoke where they had just been, but they also grew more determined not to change zones again but to wait longer for their food. People react in similar ways—for example, when we switch lines at the grocery store and realize we've made the wrong choice. The researchers detected patterns of activity in the brains of the rats similar to patterns in our brains when we mentally replay our predicament. That's what makes regret different from disappointment. The latter kicks in when we don't get what we were hoping for. In contrast, regret kicks in when we also realize there could have been a better outcome. And researchers Adam P. Steiner and David Redish discovered that rats can clearly do that.[48]

If rats exhibit these feelings, doesn't it seem logical to look more closely for such feelings in dogs? After all, almost every dog owner knows that dogs are sorry when they behave badly, which they demonstrate by pulling a typical pitiful puppy-dog face when they're told off. Our Münsterländer, Maxi, knew very well when she had done something wrong and I scolded her. She would put her head down and look up at me, as though it was all terribly distressing and she was asking me for forgiveness. And exactly this behavior has been put to the test by researchers. Bonnie Beaver from Texas A&M College of Veterinary Medicine came to the following conclusion: dogs give their owners the typical puppy-dog look because they learn what their owners expect when they scold, which means the dogs are reacting to the scolding and not to their guilty conscience.[49] Alexandra Horowitz from Barnard College in New York came to the same conclusion. To do this, she asked fourteen dog owners to each leave their dog in a room with a bowl full of treats, after giving the dog a strict warning not to touch anything. Some owners were then misinformed about whether their dogs had eaten the treats or not. The result: although some of the dogs obeyed the order, all of the dogs made the puppy-dog face as soon as they were scolded.[50] And yet that doesn't necessarily mean that dogs pretend to feel sorry for their actions. If the scolding comes right after the deed, dogs connect their owner's reaction with their behavior, and then their look probably really does express the remorse we ascribe to them.

Let's come back once again to the feeling of fairness, for that definitely exists in the animal kingdom, and not just among horses. If you live in a social group, things need to be fair. According to the dictionary definition of the term "justice," every member of a community should be treated equally. If they aren't, resentment quickly bubbles to the surface and, if this resentment is constantly fed, it can lead to violence. In human communities, laws are supposed to protect everyone's interests. However, emotions such as shame when we behave badly and happiness when we behave well are considerably more important than the law when it comes to our daily dealings with each other. How else could fairness function within the privacy of our own four walls, within our own family?

I've already written about how our horses feel ashamed and have an innate sense of fairness. Of course, this is a personal observation I've made and not the proven result of a scientifically controlled experiment, but such experiments do exist for dogs. To prove that a sense of fairness exists in dogs, Friederike Range's team at the University of Veterinary Medicine in Vienna had two dogs that knew each other sit side by side. All the dogs had to do was follow a simple command: "Give me your paw." After that, there was a reward, which could vary greatly. Sometimes it was a piece of sausage, sometimes only a piece of bread, and sometimes it was—nothing at all. As long as the rules of the game were the same for both dogs, everything was fine, and the dogs were happy to play along. To make the dogs feel envious, as the experiment progressed, the dogs were rewarded most unfairly. If both offered up a paw, only one got a reward and the other didn't. The more stringent tests rewarded one dog with sausage while the other got nothing, even though it had offered its paw like a good dog. The unfair gift of food to the other dog was eyed suspiciously. It didn't matter whether the other animal got the tastier treat with or without doing anything; at some point the dog that was being unfairly treated got fed up and refused to cooperate. But if the dog was sitting alone and wasn't able to compare itself with its companion, then it didn't object and continued to cooperate even when there was no reward. Before this experiment, researchers had observed such feelings of envy and unfairness only in primates.[51]

Ravens also have a strong sense of right and wrong. Researchers discovered this while running experiments

designed to explore cooperation and tool use. In these experiments, a small board with two pieces of cheese was put out
behind bars. The cheese was attached to a cord, and the ends
of the cord were threaded through the bars to the two ravens.
The birds could only pull the treats within reach if they both
pulled carefully on the ends of the cord at the same time. It
didn't take these clever birds long to catch on, and the test
worked particularly well with couples that liked each other.
But what happened with some other teams was that after successfully pulling the cheese closer, one of the ravens snapped
up both pieces. The bird that ended up with nothing noticed
this and from then on refused to work with its greedy colleague. Selfish individuals, it seems, aren't popular in the bird
world, either.[52]

21

EMPATHY

THE COMMONEST MAMMALS in the wood are also among the smallest representatives of this class of vertebrates: wood mice. They are pretty little things but, because they are so small, they are difficult to observe and therefore many woodland walkers are not that interested in them. I didn't realize how many of these tiny creatures are scurrying around in the underbrush until I had to stand around for a long time in one spot in the woods waiting for an appointment with someone interested in our woodland burial ground. Wood mice are omnivores, and they spend their summers under old beech trees living the good life. There are buds, insects, and other small animals aplenty, so they can take it easy and concentrate on bringing up their offspring. But then winter approaches. To avoid the worst of the cold, they set up their living quarters at the foot of mighty trees, usually where multiple roots cover the forest floor. The roots create natural hollows and all the mice need to do is enlarge them a bit for their purposes. Wood mice are social creatures, so a number of them usually share the same space.

When there's snow on the ground, I can sometimes spot the traces of a drama that played out among the roots. A trail of small paw prints leads to the trunk of a beech—a marten on the move. And martens love mice for breakfast. The tracks lead to a root hollow, and now I can clearly see signs of a whole lot of scraping and scratching. Not only did the marten non-chalantly dig up the hidden hoard of mouse food, but it might even have dug up one of the mice. What must that have been like for the other mice? Were they simply afraid of the marten or did they also realize that its activities caused one of their own to suffer?

Clearly they were aware of their fellow mouse's suffering, as researchers at McGill University in Montreal have discovered. The researchers observed evidence of empathy in the tiny mammals, the first non-primates in which such emotions have been documented. The experiments themselves, however, were anything but empathetic. The researchers caused painful injuries by injecting acid into their tiny paws, or, in another experiment, pressed these sensitive body parts onto hot surfaces. If the mice had observed other mice experiencing similar torture, they experienced considerably more pain than if they went into the experiment unprepared. On the flip side, the presence of another, less-traumatized mouse made it easier for the test subject to endure the pain. What was important was how long the mice had known each other.

There were clear effects of empathy if the animals had been together for more than fourteen days, which is typical for wild wood mice in Central European woodlands.

But how do mice communicate amongst themselves? How do they know another mouse is suffering and experiencing a private hell? To find this out, the researchers blocked their senses one after the other: sight, hearing, smell, and taste. And although mice like to communicate using smell and make shrill ultrasonic calls when alarmed, surprisingly enough, in the case of empathy, it is the sight of suffering companions that triggers their response.[53] So when a marten fishes a wood mouse out from a cozy root hollow in winter, the other mice must endure their own few seconds of terror. We don't yet know how long feelings of empathy last. So when I come across the marten's tracks in the snow, I have no idea whether the sympathetic responses of the wood mice and the upset they cause are still distressing the tiny inhabitants below.

But how does empathy work with mice that have only recently arrived and therefore are not yet integrated into the group? Clearly, it is felt considerably less intensely, and in this respect mice are surprisingly no different from people, which the researchers at McGill University also discovered. They compared the empathetic behavior of students and mice, and concluded that empathy for family members and friends is much more pronounced than empathy for strangers. The reason was the same for all the experimental subjects—stress. Stressed individuals are less affected by the suffering of others. The strangers themselves are often the cause of this stress, and the sight of them releases the hormone cortisol. To verify this, researchers carried out another experiment, this time using a drug that blocked the production of cortisol in both students and mice, and feelings of empathy increased again.[54]

Even domestic pigs put in another appearance when it comes to empathy. In this case, I'm thinking of the experiments of Dutch scientists at Wageningen University and Research Centre in charge of the experimental pens at the Swine Innovation Centre in Sterksel, where they played classical music to the pigs. Don't worry, the researchers weren't trying to find out if pigs are fond of Bach. Rather, they got the pigs to connect the music to small rewards, such as chocolate-covered raisins hidden in the straw. Over time, the pigs in the experimental group came to associate music with particular emotions. And now things got interesting, for other pigs were added that had never heard such sounds and therefore had no idea what they meant. Despite this, they experienced the same emotions the musical pigs experienced. If the musical pigs were happy, the newcomers also played and jumped around; in contrast, if the musical pigs were so scared that they urinated on themselves, the newcomers caught the feeling and exhibited the same behavior. Pigs clearly can experience empathy. They can pick up on the emotions other pigs are feeling and experience those feelings themselves—a classic expression of empathy.[55]

And what do things look like, emotionally speaking, between different species? It is clear that we react to the suffering of other species. Otherwise, why would we be so disgusted by photographs of naked, bleeding chickens in dark battery cages or apes with exposed brains attached to research equipment? A particularly moving example that animals, too, are capable of empathy across species lines comes from the Budapest Zoo. Zoo visitor Aleksander Medveš was filming the brown bear in its enclosure when suddenly, a crow fell into the moat. The bird began to weaken as it thrashed about and was in danger of drowning, when the bear intervened. It carefully took one of the bird's feathers in its mouth and pulled the bird back to land. The bird lay there as though petrified before it pulled itself together. The bear took no more notice of this fresh morsel of meat, which was definitely a potential prey item. Instead, it turned its attention once more to its meal of vegetables.[56] Accident? Why would the bear do such a thing when it clearly had nothing to do with either the urge to eat or the urge to play?

Perhaps it would help answer the question of whether a species is capable of empathy if, in addition to direct observation, we were to take a look inside the brain. To locate empathy, researchers check for the presence of mirror neurons. These specialized cells were discovered in 1992, and they have an unusual quality. When you engage in a particular behavior, normal nerve cells in your body fire electrical impulses. Mirror neurons, in contrast, become active when someone else engages in the same behavior. That is to say, mirror neurons react as though it's your body that is affected. A classic example is yawning. When your partner opens his or her mouth to yawn, you feel the need to yawn as well. (It's more enjoyable, of course, when you allow yourself to be infected by someone else's smile.) More serious situations make this even clearer. If a member of your family cuts their finger, you suffer along with that person as though you had injured yourself, because similar nerve cells in your brain respond. However, these mirror neurons function only if they've been trained early in life. People who have loving parents or other caregivers practice mirroring emotions and strengthen these neurons from an early age, while the capacity to feel empathy wastes away in people who are denied early exposure to this skill.[57]

Mirror neurons, then, are the hardware of empathy, and so what would make more sense than to look and see which animals possess these cells? And this is exactly where modern research now stands: all researchers know so far is that apes possess mirror neurons. We still need to test to see which other species are like us in this respect. Scientists often publicly speculate that we can probably expect surprises here, too. They assume that all animals that live in herds or large groups possess similar brain mechanisms, because social units function only if individuals can see things from the perspective of others in the group and feel what they are feeling. I can just see the goldfish in the chapter "Anybody Home?" waving its fin at us again. As an animal that travels around in a tightly knit group, it's on board with this idea—or at least swimming alongside the boat.

22

ALTRUISM

C AN ANIMALS ACT selflessly? Selflessness is the opposite of selfishness, a characteristic that, in evolutionary terms (only the fittest/best survive) is not inherently negative. However, if you live in a community, a certain level of selflessness is required for it to function, at least when this characteristic is defined in a way that is not necessarily tied to free will. Under this definition, many animals act selflessly—even bacteria. For instance, individual bacteria that are resistant to antibiotics release indole, a substance that serves as an alarm signal. Immediately, all the other bacteria in the area take protective measures. And this means that even those that have not mutated to become resistant to antibiotics are able to survive.[58] This is a clear case of selflessness, but, according to current scientific opinion at least, it's doubtful that free will comes into play.

As far as I'm concerned, altruism is only meaningful when you have to make a real choice—when you have to consciously and actively sacrifice something in order to help another. In the final analysis, we can't be sure when this is the case with animals, but we can get closer to the issue by starting with the more intelligent ones. Birds belong in this category, and we can observe them behaving altruistically all the time. When an enemy approaches, for example, then the first great tit to notice the danger gives a warning call. Now all the other tits can fly away to safety. The bird that makes the call, however, has put itself in danger by drawing attention to itself. Of course, it can also try to fly to safety, but it is now highly likely that the attacker will catch it instead of one of the other tits. Why would the tit run such a risk? In strictly evolutionary terms, it makes no sense, because as far as the species itself is concerned, it makes absolutely no difference which bird is eaten. However, in the long term, altruism means not only giving but also taking, and that can mean future advantages for sympathetic and generous individuals, as Gerald G. Carter and Gerald S. Wilkinson of the University of Maryland observed in, of all things, vampire bats.

These South American bats come out at night and bite cattle and other mammals so they can lick up the blood that flows from the wounds. If they're going to eat their fill, however, they need experience and luck when it comes to finding cattle and making sure their victims don't move. Unlucky or inexperienced bats often go hungry, but only until their well-fed colleagues return to the cave. Here, the successful bats regurgitate part of their meal of blood for their less fortunate cave mates, so they all get to share in the meal. And I mean all. Surprisingly enough, it is not only close family members that get to eat, but also other bats not even distantly related to the one dispensing food.

Why do the bats do this? From an evolutionary standpoint, surely it is the strongest that survive, and giving handouts makes animals weaker rather than stronger. Finding food takes energy, and animals that feed others as well must use more of it while exposing themselves to danger more often. In addition, some members of the community could exploit selfless bats and permanently take advantage of their services. But it's not like that, as the two American researchers discovered. You see, the bats recognize one another and know exactly which of their acquaintances are generous and which are not. Those that exhibit especially altruistic traits are the first to be looked after if they themselves ever run into a string of bad luck.[59]

Does that mean that altruism is selfish? In evolutionary terms, certainly, because the individuals that show these traits have a higher chance of survival in the long term. But there is something else we can learn from the scientists' observations. Clearly the bats have a choice—free will—and they can decide to share or not to share. If that wasn't the case, there surely would be no need for the complicated social network of mutual recognition, attributing particular traits to particular individuals, and the behavior this gives rise to. Altruism could simply be genetically fixed as another reflex so there would no longer be any recognizable character differences between the bats. However, selflessness is meaningful only if it happens of the individual's own free will, and vampire bats clearly exercise their ability to make this choice.

23

UPBRINGING

JUST LIKE HUMAN children, animal children need to be taught how to master the rules of adulthood. Miriam and I were to learn just how important this is when we bought our small herd of goats. The owner of the dairy in the neighboring village sells only kids, because he needs the milk from the mother goats to make cheese. That means the baby goats have two alternatives: either they end up at the butcher's as meat or they are sold to hobby farms. Our starter herd of four was lucky enough to come to our pasture as a small group. No sooner had they stepped into their fenced-in enclosure than the first little goat panicked, jumped the fence, and disappeared into the forest about 800 yards away. We assumed we would never see it again. After all, how would it know where its new home was? Under normal circumstances, its mother would have been by its side. She would have bleated to calm it down until it felt safe. As it was, it didn't have a support system. None at all? What about the other three kids? Although they were all part of the same herd, they clearly couldn't create anything like a safe space for each other. And so the problems continued.

Bärli (the brown runaway) returned, but then the other members of the rascally band took to hanging out outside the fence, and we broke out into a sweat when we had to round them up. Our only hope was that their behavior would improve after the first young were born. And indeed, as soon as the goats had their first kids, they became calmer and stayed obediently on their designated patch of pasture. Their sons and daughters didn't give us any trouble at all, because they learned from their mother how good goats live their lives out in the field. If a kid stepped too far out of line, it was first called to order with a sharp bleat and, if that was not enough, it was then at the receiving end of a swift butt from its mother's horns. Not a single one of this second generation of goats has ever jumped the fence, and the escape artist Bärli is now our most well-behaved and affectionate goat, regal and relaxed. Of course, getting older helped: Bärli is heavier and therefore a bit less agile than she used to be, but she is also calmer in herself. Having children has certainly given her self-confidence, and she has now risen to become the leader of the herd, which surely brings more peace to her life.

Doesn't that all sound perfectly normal and natural? I think so, too. However, if you believe that animal behavior is instinctive and follows genetically fixed programming, the whole situation looks somewhat different. If that were the case, learning would be unnecessary, because every situation would activate an appropriate behavioral response. But this is certainly not what happens, as millions of pet owners can attest. For example, our dogs are not allowed in the kitchen, and they learn quickly that the kitchen is off limits when we say "No!" in a very particular tone of voice. For the rest of their lives, they obey the rule, even though it's a rule that makes no sense at all in the animal world.

But let's take another look at teaching wild animals about life out in the woods, and let's start with the smallest: insects. If insects don't grow up in a colony like bees or their relatives, the ants or wasps, then the young tykes are on their own. There is no one to warn them of the dangers of everyday life, and they have to learn about them all by themselves. It's no wonder that most insect young are eaten by birds or other enemies, and perhaps this parentless learning curve is the main reason insects have so many young. Mice also reproduce rapidly, but considerably less rapidly than these tiny creatures. Field mice, for example, have offspring every four weeks, which, in turn, can bear young when they are just two weeks old. However, the little rodents don't walk off and abandon their young to the world. They teach them how to interact with their environment and how to find food. Scientists have researched just how specific this training can be in the case of house mice, which are common where I live. But the research took place far, far away on Gough Island in the raging South Atlantic, thousands of miles from the nearest mainland.

Seabirds such as the enormous albatrosses breed here in complete isolation. That is, they did until one day seafarers discovered the island and inadvertently released house mice that had stowed away on their ships. The mice did there what mice do here. They dug holes, ate roots and blades of grass, and multiplied magnificently. But then, one day, one of them suddenly got a taste for meat. It must have found out how to kill albatross chicks, which, quite apart from the savagery of the act, is no easy feat because the chicks are around two hundred times larger than their attackers. The mice quickly learnt that a large number of them had to keep biting a chick until it bled to death. Especially brutal mice even began to eat the fluffy balls of down while they were still alive.

But back to animal school. The researchers noticed that, for years, chicks were hunted only in particular parts of the island. Clearly, mouse parents demonstrated the technique to their children, passing the skill on to the next generation, while other mice in other areas knew nothing about this hunting strategy. This handing down of hunting strategies is found among many larger mammals, as well, such as wolves. Moreover, young wild boar and deer are also taught by example to follow the paths their family groups have been navigating safely for decades to move from summer to winter feeding grounds. And that is why such trails are often well trodden and hard as concrete from long use. I can tell you that animals that learn from older generations avoid an early death. Unfortunately, I cannot tell you whether school in the animal world is any more enjoyable than school in ours.

24

GETTING RID OF THE KIDS

I T WAS CLEAR to us, as it is clear to most parents: one day our children were going to have to stand on their own two feet. We brought them up to be independent from an early age, and Nature—more specifically, hormones—did the rest. Although puberty entered our house with little fanfare, at this stage in life, differences of opinion often led both sides to think that a parting of ways at some point in the future might be desirable. The educational system played its part, as well. After graduating high school, it's time for college. As there is no college close to our isolated forest lodge, that meant both our children had to move to Bonn, about 30 miles away. Incidentally, the parent-child relationship in our family improved markedly at this juncture, because we were no longer constantly getting on each other's nerves. And what do animals do? With mammals and birds, at least, there is a similarly tight bond between generations that has to be loosened at some point. And most animals have an additional problem: because the majority of them don't have families in the human sense, within a year at most the adolescents have to make way for new babies. So how do animals get their children to move on?

One approach leaves a bad taste in the youngsters' mouth. And I mean that literally. We've experienced this ourselves with our milk goats. If a mother loses her kids in spring as a result of some misfortune, we have to take matters into our own hands and milk her. If we didn't do this, her engorged udder could get inflamed, which would be very painful. To say nothing of the fact that we then get delicious milk that we either pour over our cereal or process into cheese. Did I say delicious? Well, in the first few weeks, that's exactly what it is. It tastes smooth and creamy, and it's hard to tell from good cow's milk. However, as spring progresses, the milk tastes increasingly bitter. Eventually, none of us want to drink it anymore. This is when we increase the time between milkings, and the goat's milk gradually dries up. It doesn't matter whether it's us or the kids drinking the milk—the taste makes the udder unappetizing, and the young goats begin to turn to grass and other things green. That takes pressure off the mother, as the kids no longer depend on her for food. In addition, she allows them only a few seconds at the teat before she gets irritated, lifts her leg, and butts them away. Right on schedule, by fall—and that means mating season—she will have had time to top up her physical reserves both for herself and for the young she will bear the following year.

Although bees don't want to get rid of their children at the end of the summer, they do want to get rid of their men. The drones—gentle, big-eyed creatures without a sting—mooch about the hive all spring and summer. They don't look for flowers. They don't help dry nectar and process it into honey. They don't feed or look after the baby bees. Nope. They enjoy the good life, getting fed by the worker bees and occasionally flying out of the hive to check if there might be a queen hanging out nearby who is ready for their attentions. If they spot one, they're on her trail right away, but only a lucky few manage to mate with her on the fly. The unsuccessful suitors buzz their way back to the hive, where they allow themselves to be consoled with a sugary meal. They could live like that forever. However, as summer runs its course, so does the worker bees' patience with the moochers. The young queen mated long ago, and her sisters, who have left the hive in swarms, have also been impregnated. Winter is slowly approaching and the precious provisions in the hive have to feed a few thousand overwintering bees—workers that have managed to survive to a ripe old age—and the queen.

Nothing has been put aside for the burdensome drones, and now a hateful chapter in these insects' life cycle begins. During the late summer drone massacre, the once-pampered little males are rudely grabbed and unceremoniously shown the door. Resistance is useless, even though the drones use their legs to brace themselves against removal. Clearly they don't like this one bit, and their senses are all on high alert. But any drone that puts up too much of a fight is simply stung to death. The workers show no mercy. Any drone left alive eventually dies an agonizing death from starvation or quickly finds itself in the stomach of an equally hungry tit.

25

ONCE WILD, FOREVER WILD

A FEW YEARS AGO, I got a telephone call from the next village. A worried woman told me she had a fawn in her house and didn't know what she should do. When I asked her questions, I discovered that her children had brought the animal back to the house after they had been playing with it in the woods. Damn! However well-meaning that playful gesture might have been, it was a catastrophe for the young animal. In their first few weeks of life, fawns are left lying alone in the undergrowth or tall grass, because this is safest for both mother and fawn. A doe with a fawn moves slowly because she has to keep waiting for the youngster. Often the fawn has not yet experienced how serious life can be, and it dawdles behind mom—an ideal target for wolves or lynx. These predators can spot the pair from a long way off and easily grab a meal. That's why mother deer prefer to separate themselves from their little darlings for the first three to four weeks and leave them in a safe place. It is almost impossible to sniff out a fawn. Because they smell of hardly anything at all, their scent doesn't alert predators to their presence. The doe comes by for a quick visit every once in a while to nurse her fawn, and then she takes off again right away. That gives her more time to feed on nutritious buds and new shoots instead of having to worry about and watch out for the little one all the time.

When someone who has no idea what's going on finds a fawn lying there all alone and very still, their almost instinctive reaction is to take care of it. After all, it's hard to imagine what an abandoned human baby would go through if someone just put it down somewhere and then disappeared. And so, time and again, "helpers" spontaneously step in and take the supposedly orphaned fawn back home with them. However, they have no idea what to do next, and they call an expert. It's usually about this time that they realize bringing the fawn home was a huge mistake, but by then it's too late to fix the damage. The fawn now smells of people, and it can't be returned to the wood and its mother, because she will no longer recognize her child. Bottle feeding a fawn is hard work and, at least in the case of male fawns, it is also risky, as we shall see in due course. I find deer a perfect example of how mother love can be expressed in very different ways. Most mammals are like us and seek constant close contact with their offspring; those that behave differently are not heartless but simply adapted to different life situations. Fawns feel perfectly safe in the first few weeks of life even without constant contact with their mother. Their behavior changes only when they are capable of bounding along behind her. Then they stay close to the doe, rarely straying more than 20 yards away from her.

Unfortunately, in these modern times, typical fawn behavior in the first few weeks of life has other, far more tragic consequences. When danger threatens, a fawn hunkers down, because it knows instinctively that it is extremely difficult to pick up its scent. But often what is threatening it is not a wolf or a hungry wild boar looking for a succulent morsel of meat. What is bearing down on it is a tractor fitted with an enormous mower that is swiftly cutting down grass on acres of land. And so the crouching fawn ends up under the blades, and in the best-case scenario it is killed immediately. Often, however, it stands up just before the blades reach it so that its legs are cut off along with the grass. One way to avoid this would be to have someone walk through the area the evening before with a dog along as an extra signal that this is a dangerous place to be. Then the doe will encourage her fawn to accompany her to a safer place outside the meadow. However, there is often neither the time nor the staff for such rescue operations.

Another example showing that wild animals are not suited to domestic life or cuddling is the European wildcat. By 1990, it was almost wiped out. Only about four hundred animals survived in the central uplands in western Germany, and there was a small remnant population of about two hundred individuals in the Scottish highlands. The forest I manage in the Eifel was one of the last remaining refuges for the wildcats, and so I was frequently able to observe one of these diminutive feline carnivores. Since then, their situation has improved considerably. Thanks to conservation and reintroduction efforts, thousands of wildcats now roam once again through the wooded landscapes of Central Europe.

The identifying features of a European wildcat are clear. It's about the size of a powerful-looking housecat. Its dense coat is tiger striped with a touch of ochre. Its bushy tail is ringed and has a black tip. The problem is that this is what domestic tabby cats look like as well, even though they are not related to the wild species. Positive identification is possible only by measuring brain size or gut length, or sending a hair sample off for genetic testing. And, of course, forest visitors don't usually have any of these testing methods at their disposal. Despite that, there are a few clues. Domestic cats have—how shall I say it?—grown a bit soft, and they stalk through the landscape only at warmer times of the year and no farther than a mile or so from their home. As soon as it gets cold and wet in winter, their desire for adventure and the radius of their activity diminish, and they usually don't stray more than 500 yards in any one direction, because chilly housecats want to be able to return quickly to the warmth of their home. Of necessity, wildcats are tougher. They don't hibernate or dial down their metabolism, and they have to hunt mice even when there's snow on the ground. Tabby cats out in the snow, miles from the nearest village, are therefore certainly wild and free.

Since Roman times, housecats introduced from southern Europe have vastly outnumbered wildcats. Why, then, have the latter not been extirpated through cross-breeding? The existence of hybrids proves that the two species do interbreed. However, that is the exception. When the two species meet, the tame version always comes off worse, because wildcats are quick to live up to their names. And this brings us to the question whether wildcats make good pets. In rural areas, there must have been many times when individual animals attached themselves to people, and no doubt this is still the case. After all, there are enough animal lovers today who put out food, and the birds that flock to winter feeders show that wild animals become less shy around people over time.

Recently in my own village, I found out what happens when a baby wildcat grows up cared for by humans. A jogger had spied a youngster next to an isolated path in the woodland I manage. He resisted the urge to scoop up the clearly helpless animal and take it back home with him, and so at first he just watched it. A few days later, he came back to the same spot, and found the mewing little fur ball still sitting beside the trail. Clearly, for whatever reason, its mother had gone missing, and left to its own devices, the baby cat would have died. This time, the jogger picked it up carefully and brought it back home with him. He checked with a wildcat recovery center to see how best to care for the animal, and the Senckenberg Research Institute in Frankfurt confirmed, by testing a hair, that this kitten was 100 percent pure wildcat.

Because of their shorter intestines, wildcats cannot tolerate cat food, and so the little wild child was fed meat. It wasn't long before the family could no longer get close to it at feeding time, because the kitten went on the attack right away. Apart from that, it stayed by the family's side when they went for walks through the meadows and it seemed it might be getting tame after all. But the cat became even more aggressive and began scaring the older housecat until it was impossible to keep it any longer. The family eventually dropped it off at a wildlife rehabilitation center in the Westerwald.

The story shows that many species do not lose their wildness and therefore are not suited for life in the care of humans. There is good reason why every domestic animal today is the result of a long breeding process. And anyone who is just itching to adopt a wild animal despite the fact that they are ill-suited for domestication will run afoul of the law, as well. Depending on the country, the laws governing conservation and hunting are very strict and state that wild animals can only be kept in exceptional cases with official approval.

And yet some people try to make the impossible possible, especially in the case of the wolf, which is unfortunate because it was difficult enough to gain sufficient support for its return to Central Europe without any further complications. The wolf poses no danger to humans, because it takes absolutely no interest in us. However, when we imprison it against its will, the situation is very different. So here's the thing. Not only is keeping a wolf against the law, but just like the wildcat, it remains a wild animal at heart. The next thing people think of is to just cross the wolf with a large dog, such as a husky, to get an animal that has the appearance of a wolf but the tractable nature of a domestic dog. But that is illegal, as well, which has led to a black market for these animals, which are imported from the US or Eastern Europe.[60] The high proportion of wolf blood in these wolf-dog hybrids ensures that they are never really tame, and they are stressed when they have to endure living with people. Such forced proximity is inevitably dangerous because stress leads to aggression.

Dr. Kathryn Lord at the University of Massachusetts researched why wolves, which are, after all, highly social animals, are so much more difficult to keep than dogs. Her results show that the answer lies in the way the pups are socialized. Baby wolves are up and about at the age of two weeks, before they've even opened their eyes. They can't even hear at this stage. Their hearing develops and becomes functional only after four weeks. So for the next couple of weeks, they feel their way around their mother, constantly learning about their world while they are still deaf and blind. They finally gain control of their eyes at the age of six weeks, but by then the little rascals are already familiar with the smells and sounds of their family and their surroundings, and they are firmly integrated into the social life of the pack.

In contrast, dogs are late bloomers, and that's just what they need to be. They mustn't bond too soon with other pack members, because when it comes down to it, their relationship with a person will be the most important relationship of their life. Thousands of years of breeding have delayed the socialization phase in dogs, and today it starts when they are four weeks old. With both wolves and dogs, the formative period lasts only four weeks. While not all the wolf pups' senses are fully developed at this important time, puppies explore their environment equipped with their full sensory repertoire—and in the final days of this phase of their life, people are part of their environment. This means that whereas dogs basically feel most at home in our company, wolves retain a certain distrust of us all their life.[61] Wolf-dog hybrids clearly retain this fundamental feeling.

In comparison with a fawn, however, a wolf-dog hybrid is harmless. A fawn? Not all fawns, but males are indeed potentially lethal for their owners. For in less than a year, the adorable dappled baby grows into a mature buck. Deer are loners and do not tolerate any competition in their territory. The affectionate connection established between owner and deer during the rearing process fades away, and because its caregiver is clearly another deer (at least in the eyes of the buck), it must be a rival. And so it must be forcibly driven off. Any owners who cannot sidestep gracefully like the buck's natural competitors will find themselves blindsided by a sharp-horned battering ram. Such behavior is not the exception but the norm. Even when the animals are released back into the wild, the danger remains. After all, deer have memories, too, and they don't always avoid people later in life. There was a report in a 2013 issue of the *Schwarzwälder Bote* of a buck that attacked and wounded two women in the evening hours on a playing field in the village of Waldmössingen. It turned out that it had been hand reared the previous year.[62]

26

SNIPE MESS

AS I HAVE already mentioned in the chapter "Shame and Regret," our horses Zipy and Bridgi are fed a portion of concentrated feed at noon. The energy-rich grain is meant to beef them up a bit, especially our older horse, Zipy. Clearly, horses don't chew their food very thoroughly, because we find a fair number of undigested seeds in their droppings. And now things get rather unappetizing, because our "house crows," which are always hanging about close to the pasture, have their eye on these. They pick the horse apples apart and peck out individual oat seeds. Yummy? I think it looks pretty disgusting, and so the question is, how tasty can such excrement-coated food really be? Do animals have any sense of taste at all? They definitely do, but it is adapted to different traditional fare than our palates. (Of course, there are also differences in taste perceptions in people. Just think of the hundred-year eggs so beloved in China. To many Europeans, at least, these dark, translucent eggs conjure up visions of rot and decay rather than culinary delight.)

Our horses offer further proof of animals' sense of taste. We have to worm them two or three times a year. To do this, we squeeze a medicinal paste from a tube into their mouth. Apparently, it doesn't taste good at all, because when the two of them notice what's about to happen, they really don't want to be around us. However, the manufacturer has recently made some changes. The deworming treatment now comes in apple flavor, and horses love apples. Since then, the procedure has gone somewhat more smoothly. Dog owners are also well aware that their pets learn what they like and what they don't like. For instance, if they change food brands, their four-legged companions sometimes refuse to eat. This was never a problem with Crusty, the French bulldog, who always ate with gusto; however, unfamiliar foods exacted their price, at least for us. A short time after such a meal, a stink cloud from Crusty's rear end would waft through the air every ten minutes or so, filling the room.

And then there are rabbits, which are slightly more perverse even than crows when it comes to matters of taste. At least the birds only peck around in the fecal matter of others and restrict themselves to eating the seeds they find there, but rabbits and hares chow down on their own excrement. Though it must be said that they don't eat all their droppings indiscriminately, just the special ones. Like all herbivores, they have bacteria in their gut that help them dissolve and digest chewed grasses and other herbaceous matter. There are specialized species in the cecum, in particular, that break down greens into their component parts. However, some of the end products can only be absorbed in the small intestine, which, annoyingly enough, comes before the cecum. And so the beneficial brew slides unused through the digestive tract and inevitably ends up in the outside world once again. What better idea than to smack your lips and enjoy this product from the cecum fresh from the anus, ingest it anew, and extract valuable calories during its journey through the small intestine? The final processed waste—hard droppings—are completely ignored and clearly regarded as fecal matter.[63]

We cannot imagine eating excrement, whether it's from animals or ourselves. Or at least most of us can't. And yet, there are people who do this. And you can find them in Central Europe. They are hunters. To this day, they shoot snipe, a hunt I find as repulsive as the hunt for whales. To make it even worse, these birds have hardly any meat on them, and perhaps this is why people have developed an odd way of eating them. The parts of the bird they enjoy include "snipe mess," which is the gut along with its contents (that is, fecal matter). Finely chopped and dressed up with other ingredients, perhaps with bacon, eggs, and onions, the whole concoction is baked on bread, and there you have it—a hunter's delicacy. Despite the fact that cooking kills worm eggs and other things you find in bird feces, just thinking of such "treats" makes me completely lose my appetite.

Animals must be able to taste things in order to tell the difference between what is appropriate as food and what is not—and also what is poisonous. By all accounts, however, it seems that many other species don't taste things the way we do. For example, there's a reason the popular children's storybook character Pooh likes honey and his friend Tigger doesn't. Large carnivorous cats such as lions and tigers and even marine carnivores such as sea lions have lost their taste receptors for sweetness over the course of evolution. Clearly, sugary food is of no particular interest for these animals, and it's easy to understand why: meat doesn't taste sweet at all.[64]

It is even more difficult to compare our sense of taste with that of butterflies. Consider the swallowtail. The female lays her eggs only where her young have immediate access to the juicy leaves of suitable plants. That way, all the freshly hatched caterpillars have to do to satisfy their hunger is to munch on the leaves around them. However, the butterfly doesn't have to use her mouthparts to test each plant on her way to lay her eggs, because she can check them out with her feet. As she walks around a leaf, her feet, which are equipped with sensory hairs, can taste up to six different substances. And that is not all. The butterfly can even detect the plants' age and health.[65] Does that sound unbelievable? We, too, can detect whether something is fresh or past its prime just by tasting it. Think of an overripe banana, for example. Being able to taste a plant's health could be crucial to the survival of a butterfly's young. After all, if a plant dies before her caterpillars pupate, their dreams of metamorphosing into butterflies are over.

27

SOMETHING SPECIAL
IN THE AIR

FTER THE SENSE of taste, it seems appropriate to take a close look at the sense of smell. Animals definitely have an idea what smells good and what smells bad. This sense not only serves, as with taste, to test food but is useful in other ways, as we well know, and one of these is making oneself attractive to the opposite sex. I have to say that in fall our billy goat, Vito, demonstrates just how far humans and animals diverge when it comes to the scents we experience as pleasant. As I mentioned earlier, he uses his own perfume— his urine—to make himself attractive to the two female goats in the herd. Because of this, my wife changes her clothes and wears a cap when she visits them in the stable, because the penetrating smell doesn't only drift through the whole yard, it also infiltrates fabric and hair.

But what we find disgusting might just be due to the culture of our times. Acquired tastes might explain why deodorants were not yet in common use two hundred years ago. Out on a military campaign, Napoleon supposedly once wrote to Josephine: "I return to Paris tomorrow evening. Don't wash!" The sixteenth-century Spanish conquistadors, wanting to differentiate themselves from the fastidiously clean Moors they had just driven from the Iberian Peninsula, also had their doubts about the benefits of bathing. When the Aztecs in Mexico met these light-skinned strangers for the first time, they immediately smelled the difference between them and their own people, who used steam baths for their personal hygiene, and it was nauseating. Old, well-aged cheese is a contemporary olfactory example. You could also describe it as rotten, dried-out milk protein giving off fumes that, under other circumstances, would make people gag. I list these examples not to put people on the same odiferous level as animals that stink. Not at all. I am merely pointing out that people, too, differ greatly in what they perceive as smelly.

Dogs trump billy goats when it comes to stink. Our dog Maxi loved to roll around in fox feces, which have a particularly penetrating aroma. Fresh cow pats were another source she liked to use for special scents. For a long time, people assumed that their four-legged companions rolled in dung to disguise their scent. The theory was that this gave dogs— or at least their wild ancestors—better luck when hunting. Today, dogs and wolves are thought to use scent to relay messages—or perhaps they simply want to stand out in the pack. And clearly they don't find the smell, especially the smell of rotting meat or herbivore feces, unpleasant. Quite the opposite, in fact.[66] Does that remind you in any way of people and their relationship to perfume?

You just need to be careful if your dog rolls in, or even eats, fox or dog feces. Fox feces in particular may contain fox tapeworm eggs. These eggs are as tiny as specks of dust, and after a fecal bath, they shower down from the fur of your beloved pet—likely as not, all over the living room. You then take the place of the mouse that was the eggs' intended host. The developing larvae settle in internal organs and make the host sick to slow it down. Foxes then catch the slow-moving mice, which completes the worms' life cycle. But not, of course, if you are acting as the intermediary host instead. People get a severe infection that can be difficult to cure, depending on how far the worm infestation has spread. When your dog comes home decorated with fresh feces, this is one time when it's definitely a good idea to give it a thorough drenching in the shower.

Even though their rating system differs from ours, animals are aware not only of things that smell good but also of things that stink. Particularly when it comes to their own excrement. Herbivores are better off not browsing in places where they've deposited a pile of droppings, because practically every deer, goat, or cow has worms, and a correspondingly large parasitic legacy can be found in feces. Take lungworms, for example. For every ounce of feces, there can be up to twenty thousand eggs, which are then re-ingested when the animals browse. Because massive worm infestations physically weaken the host, infested herbivores are the first to fall to lynx and wolves.

So it is only logical that an animal perceives its own excrement as absolutely revolting to ensure it gives it a wide berth.[67]

I believe that the smell of their own feces is just as awful for most animals as the smell of human feces is for us. Many domestic animals indicate that this is indeed the case. In the pasture, our horses retreat to a private corner that they use only for defecation. Out in the wild, thanks to horses' ability to move around freely, there would be no danger of eating in the same place too often. When we inhibit such movement, the horses make do by finding appropriate spots in the pasture that they use only for this purpose. Even Blacky, Hazel, Emma, and Oskar, our rabbits, designate a toilet area in their hutch and run where they do their main business. The only place this doesn't happen is on factory farms. In these conditions, chickens and pigs even have to bed down in their feces. The only way to avoid severe worm infestations on these farms is by regular doses of medications. It's a shame the pills don't get rid of the stink at the same time.

Many animals are also as embarrassed about defecation as we are. When Crusty, the bulldog, was on a leash, he would pull away from us into the underbrush if he had major business to attend to. What's more, he would turn his back to us so he didn't have to look at us. Clearly, he was embarrassed to be seen when he was squatting. Apart from the smell, it is also just as important for all animals to be clean. Like us, they feel very uncomfortable if they have feces or other dirt stuck to them. It's probably the reaction of their fellow animals that reinforces their feeling of discomfort. An animal with a dirty backside signals that it's probably ill and that's why it has diarrhea. Other animals don't want to catch something nasty and certainly don't want to mate with a soiled partner.

And that's why, like us, animals take pains to clean themselves. Although it must be said that "clean" means something different to animals than it does to us. Wild boar, for instance, love to cool down in summer, and they do this by enjoying a good wallow in a mud hole. Grunting and twitching their tails, they take their time rootling around, stirring up the mud, and then settling down for a wallow. By the time the procedure is over, the pig is covered with a layer of beige-colored mud. And yet the pigs don't feel dirty. And why should they? Isn't the human equivalent an expensive spa treatment with a pack of mud or peat? The wild boar feel the same—all fresh and clean. And there's a reason for that. As the mud coating dries, many parasites such as ticks and fleas are baked into it. Once the crust has hardened, the pigs rub it off on specially selected rubbing trees. They use the same trees or stumps for many years, and over time, they are rubbed smooth. Afterward, the pigs are free not only of the little animals plaguing them but also of old bristles and hairs, which can be just as itchy. It's the same for our horses. They, too, like to roll around on the ground, especially when they're losing one season's coat and replacing it with the next. Depending on the weather, they can end up with mud all over them—just mud, mind you, not feces.

28

COMFORT

THE LANDSCAPE IN Central Europe, as in many other places in the world, is a strange patchwork, at least when seen from the point of view of a wild animal. Wide open areas unbroken by roads or settlements are a thing of the past, and you couldn't lose yourself in the wilderness anymore even if you tried. For even our most natural ecosystems, our woodlands, are not what they used to be. In Germany, there are now 20 miles of logging roads per square mile of wood so that logging trucks can reach the remotest areas. Chances are that on a cross-country trek, you won't get more than a hundred yards before coming upon a trail. This means your adventure will consist primarily in choosing the wrong fork in the road.

Trails have distinct disadvantages for the natural world. The once-loose soil is hugely compacted, and the tiny creatures that used to live in its subterranean layers have all suffocated. In addition, trails act like dams to block the flow of water, and this interference should not be underestimated. Numerous streams flow underground, and in many cases, they are blocked or diverted by the compacted soil under trails. This means some sections of woodland are transformed into parcels of swampland on which many trees fail to thrive because their roots suffocate in the foul brew. Woodland trails also pose substantial obstacles to light-averse ground beetles. The beetles, which lost the ability to fly long ago, don't dare leave the darkness under the trees to cross paths flooded with light. Confined to a small territory surrounded by trails, they can no longer exchange genetic material with their neighbors.

But trails don't have to be a hindrance to animals. Like us, deer and wild boar walk, and they prefer not to struggle through the undergrowth. A stroll through wet grass or bushes in rainy weather is unpleasant, which is why four-legged creatures are happy to use our nice, smooth cross-country game routes. For that's how game see our roads and trails: as cross-country game routes provided by people. The animals find it much easier to walk along them, as you can see from the numerous tracks pressed into soft spots on the surface. Where people aren't there to help out, the animals make these cross-country routes for themselves. They are, mind you, considerably narrower, measuring only the width of the animal. There's no ordered plan. For example, one day the lead sow in a herd of wild boar finds her way through the underbrush. The other boar follow, and the grass and other greenery are soon trampled down. Next time the pigs pass, there are still traces of this faint track, and it's a little easier to walk along it. Over the course of time and after years of use, the path looks like a footpath trampled down by people: all the vegetation has been crushed by trotters and the path shows up as a narrow strip of bare earth. Knowledge of these easily walkable cross-country routes is passed down from generation to generation, unless humans come along and obstruct them.

One day early in my work managing the forest, I had a fence built around a new planting of oak trees. There were way too many deer, and they would have been only too happy to eat the seedlings' juicy shoots, and so I had to protect the tiny trees. Later on, I figured out that the fence had interrupted a well-established long-distance route used by these large herbivores and had forced them to find ways around it. This led to many dangerous incidents with cars, because deer began popping up in unexpected places. The fence has since been removed, and the deer have returned to using their traditional path.

Incidentally, people form trails the same way animals do. I observed the process in the part of the forest I manage that has been set aside for burials—our "final forest resting place." This is where we lease ancient beeches to serve as living grave markers, and funerals are held when urns are buried there. The burial ground saves the thousand-year-old woodland from the axe. To keep the area as natural as possible, the forest service decided not to create any new roads or paths. Nonetheless, a few beaten paths have appeared in places where it is especially easy to walk through the trees and their million-strong offspring. Rain helps form the paths. When a front of bad weather passes over the wood, the leaves of young beeches drip with moisture for a long while after. No one wants to brush past the leaves because their pant legs would get soaked in seconds. And so people seek out routes where they can stay at least partly dry, and people coming later do the same thing, following these faint traces of a path. That's fine with me, because it means the impact of visitors is confined to a small fraction of the forest floor.

I have to add that the animals' cross-country routes also offer disadvantages to those using them, because the heavy foot traffic attracts uninvited guests. More prevalent than the predators that lurk along the paths waiting to make a meal of careless travelers are small critters waiting for their next meal: ticks. Ticks are members of the mite family, and they depend on meals of blood. Because they move very slowly, they have to wait for their victims to come to them. And where better to wait than along a well-traveled path? Here, ticks cling to stalks of grass, twigs, or leaves hanging within range of the backs of deer or wild boar, waiting for the scent of mammalian breath or sweat and the vibrations caused by approaching hoofs or trotters. So, if you're wandering through the woods in summer, it's best to avoid cross-country routes frequented by game. In winter, it's not a problem, because ticks are not active at low temperatures. And there are more stowaways than just ticks lying in wait along cross-country game routes. Plants, for example, are also waiting for a ride—specifically a ride for their young. That's why the tiny fruit of catchweed are equipped with barbed hooks. When an animal passes by and brushes up against the plant, it takes a portion of the catch-weed's seeds along with it, which will fall off farther along the path. Researchers have proven that plants like catchweed spread specifically along paths frequented by game.

But let's come back to getting your legs wet. You might have experienced how unpleasant this is when you are out walking. Why should it feel any different for animals? They feel freezing cold when their coats are wet, and they would prefer to stick to the easy trails. And these paths offer one more advantage: speed of travel. When branches are snapping in the bushes and an enemy is preparing to pounce on a fawn or a piglet to eat it, the herd wants to be able to run away as quickly as it can. And because thick branches and downed trees turn flight in the forest into a scramble over an obstacle course, the best place to run is along an open trail.

29

WEATHERING THE STORM

WHICH OF US voluntarily enters a forest in a storm? Lightning strikes in trees can be deadly, and pelting, cold rain isn't pleasant either. For a number of years, I offered survival training in the woodland I manage, where the participants spent a weekend in the woods equipped with nothing but a sleeping bag, a cup, and a knife. They slept there and, most importantly, foraged for food. One time when we were foraging, we were surprised by a violent storm. We had no choice but to let it beat down onto us. Apart from the soaking rain, lightning strikes nearby were a cause for concern. I bluffed my way through the storm with exaggerated calm so as not to worry the participants even more. Inside, however, I felt a bit of rising panic when a powerful strike hit only about a hundred yards away. I've often noticed after a thunderstorm that even without a direct hit by lightning, the area around a tree is still dangerous. On one occasion, not only did the tree that was ripped open die, but so did more than ten other close neighbors. In one extreme case, I might as well have been watching a knife-throwing act. The lightning strike caused so much pressure in a spruce that its wood splintered into multiple shards that flew through the air with such force that some of them impaled a nearby stump.

That day of survival training, we were rewarded with a wonderful wildlife sighting in the calm after the storm. The rain stopped abruptly and a break appeared in the clouds, allowing the sun to shine through, bright and warm. The vegetation around us started steaming, and suddenly a deer shot out into a small clearing. The animal was completely drenched and was looking for a warm spot to dry out. He felt just the same way we did, and I felt an instant moment of connection.

What's it like to be a wild animal, anyway? They have to tough it out year round in the wind and the weather, and when the cold season rolls around that can't be very pleasant. Or can it? Let's take a closer look. First of all, animal fur repels much more moisture than is commonly thought, thanks to a coating of oil—the one we're constantly washing out when we reach for the shampoo. Then, the direction the hair grows on an animal's back points the ends down so the individual hairs act like roof tiles and channel water toward the ground. Thus, the coating of hair on deer and wild boar keeps the rain off, and their skin stays nice and dry as long as a strong wind doesn't blow the rain sideways and up between the individual hairs. Older animals are well aware of this and, when forced by the weather, they change location to somewhere

well protected from the wind. Moreover, they position themselves so that their rear end is pointing toward the wind and
their head, which is more sensitive, is sheltered from it. The
only time they experience problems is when snow falls when
the temperature hovers around freezing, because then the
snowflakes melt and form rivulets that run down between the
hairs, making the animals shiver. They prefer a hard freeze.
Their thick winter coat insulates them so well that when it's
nice and cold, a layer of freshly fallen snow can lie on their
back for hours without melting.

Don't we feel the same way? Aren't we happier on a clear,
frosty day at 14 degrees Fahrenheit than in windy, rainy
weather at 40? Animals don't really feel any different; it's just
that in general they endure lower temperatures better than
we do. But even that is not absolutely true, and to illustrate
this, I have to mention another survival training weekend.
Years ago, I once held the training in winter, and the weather
on that January weekend was particularly awful. The temperature hovered around freezing, and every hour rain turned
to snow and back again. Even the firewood was so wet that
it was difficult to keep the campfire burning. I was expecting that it wouldn't take long for the participants to give up,
but after a night in cold, clammy sleeping bags, our bodies
seemed to adapt and none of us felt cold any longer; apparently we had reached the comfort level of wild animals.

In summer, apart from the warmth of the sun, there's yet
another reason for animals to abandon the thick canopy of
leaves under the trees for a small clearing after a downpour.
The leaves of beeches and oaks keep dripping for so long after

a rainstorm that there's a local saying: "In deciduous forests, it always rains twice." But the fact that deer get rained on even after the storm has passed is not the only thing that disturbs them: falling water droplets are loud. Prey animals can't hear approaching predators behind this curtain of sound, and the predators like to take advantage of this to do a little stalking. And that is why deer prefer to move to a clearing right after a cloudburst—so they can listen to see if everything is still as it should be.

The situation for small mammals such as voles is more challenging. Sometimes when I walk through our horse pasture in winter rains, I can see water bubbling up out of the vole holes on the hillside. How are the tiny rodents supposed to survive such conditions? Wet fur is much more dangerous for them than it is for large animals, because percentage wise for their body size they lose much more warmth, and besides, they have, comparatively speaking, an enormous need for calories: every day they must consume the equivalent of their body weight in food. When they get wet, their energy needs increase considerably. And because they don't hibernate, they don't get any break from the daily struggle of finding food. On the plus side, they prefer to eat the roots of grasses and other green stuff, and therefore they don't need to brave icy winter winds but can gather food from the comfort of their underground tunnels.

But what happens when water inundates their earth-works? The smart little animals have a special building plan to deal with this eventuality. For starters, there's a chute at the entrance. In case of danger, a vole can drop into this if it has to flee quickly underground. Then, the tunnels go deep, far deeper than is strictly necessary. After a short distance, they angle slightly upward, ending in a small chamber comfortably padded with soft grass. If it rains so hard that water runs into the burrows, the water collects in the deeper sections of the tunnels while the inhabitants sit comfortably high and dry. And because a multitude of tunnels weave their way through the burrows, the voles can flee if water does manage to reach their nests. However, the plan doesn't always work. When the whole pasture is under water after a heavy rain, which happens mostly in winter, some of the voles inevitably get caught and drown miserably in their underground chambers.

30

PAIN

I T WAS A cold evening in February and our goat Bärli was about to give birth. She was restless, she kept lying down, and milk was already flowing into her udder. My wife was worried. "I think it's taking too long," she insisted. "Shouldn't we call the vet to be on the safe side?" I calmed her down. "Bärli can do this on her own. Perhaps she just needs some peace and quiet. She's strong and healthy. I don't want to interfere unnecessarily in a situation like this."

Well, it would have been better if I had paid attention to Miriam and her sixth sense. The next morning, the kids were still not born and Bärli was clearly in pain. She was grinding her teeth, she didn't want to get up, and she didn't want to eat either. All these signs were highly alarming, and we lost no time calling a vet who knew our goats well. "He's on vacation," the vet who was covering for him told us, but she quickly drove out to our forest lodge. She diagnosed a breech that unfortunately had already died inside the mother. The vet carefully pulled out the dead baby and then gave Bärli medicine to get her to expel the placenta.

Our goat recovered quickly, and we even got her another kid to adopt. A goat farm in the vicinity had a quadruplet to give away. Four kids are too much for any mother goat. She has only two teats on her udder and not enough milk for so many mouths, and the farmer was happy to find a good home for one of the feisty little kids. We rubbed mucus from the dead kid over the new arrival (later to become our breeding billy goat, Vito). That might sound disgusting, but it meant that to Bärli, the new kid smelled like her own child, and she immediately allowed it to nurse. Both mother and child thrived—so there was a happy ending for this pair, at least.

But back to pain. Pain? We talked about fish in the chapter "Anybody Home?" but those reports are still considered controversial. Now we could get down to the neurological level and advance all kinds of arguments about similar triggers and signals and patterns of brain waves and hormones suggesting similar feelings. But isn't there a much easier way? After all, Bärli exhibited all the patterns of behavior you see in people who are in pain. Grinding her teeth (something goats don't usually do), loss of appetite, lying down, apathy. Don't some of these symptoms sound familiar?

We've experienced more direct evidence in our dealings with our chickens, goats, and horses. We keep all our animals behind electric fences appropriate for each to make sure they stay in the areas we've marked out for them. An electric fence might sound cruel, but it's the most practical solution. Barbed wire is out because of the chance of injury. And a wooden fence wouldn't keep the goats in for long, and the horses would gnaw away at the posts and bars over time. Just how an electric fence works—that is to say how it works on animals—is something I experience myself on a regular basis. Sometimes when I go out in the morning, lost in thought, to mark off a new section of the pasture for the horses, I forget to switch off the power first. Then a powerful shock rudely brings me back to earth, and I get annoyed with myself. For the next few days, I check multiple times to see if the pasture fence control really is shut off. An electric shock is a wake-up call for the instincts, sending them into overdrive.

And that's exactly how the fence works for animals, too. They experience how unpleasant it is to touch the fence once, maybe twice, and from then on they avoid it. So an electric fence works by means of an initial shock and then by the mere memory of the shock. Just like it does with me. And that's why I'm convinced our domestic animals experience the electric shock exactly the same way I do. And not just our domestic animals. With the chickens, the primary purpose of the electric net fencing we use is to keep foxes out, and it works very well. Farmers fence their maize fields with electrified wires to keep wild boar out. And pet owners who don't like visible fences can bury cables. When a dog or cat crosses the invisible boundary, they receive an electric shock from a special collar. It's up to the individual to decide whether that's okay or not, but the fact remains that all these creatures feel pain and instinctively react the same way—even me.

31

FEAR

A PERSON OR AN animal that doesn't know fear will not survive, because fear protects us from making deadly mistakes. Perhaps you are familiar with that queasy feeling you get when you're up somewhere high, like a viewing platform or the Eiffel Tower in Paris. I get a tingling sensation and the overwhelming desire to climb back down again as quickly as possible. That makes a great deal of sense from an evolutionary perspective. This inborn instinct prevented our forefathers from falling off high cliffs, which would have abruptly severed the generational line that leads to us.

Wild boar show us that not only do animals experience the gut-wrenching feeling of fear and recognize threats, but they can also process this information and use it for long-term planning. So let's take a little excursion to Switzerland and specifically to the canton of Geneva. In a public referendum in 1974, the people of Geneva voted in favor of a ban on hunting. Hunters are large mammals' worst enemy. And because hunters belong to the species *Homo sapiens*, animals that are hunted fear humans. That's why they prefer to spend the day in thick forests and brushy areas out of the sight of their dangerous two-legged adversaries and usually venture out into pastures and fields only at night. When hunting was outlawed in Geneva, the behavior of deer and wild boar there changed. They lost their wariness of people and now they show themselves during the day. But it's not only the Genevese wild boar that have changed their behavior. All around, including in neighboring France, hunters are blasting away. And as soon as the hunting season kicks off—especially in the fall, when the drive hunts that use packs of hounds start—the pigs become excellent swimmers. As blasts of the hunting horn reverberate through the air and gunshots begin to ring out, many boar leave the French shore and swim across the Rhône river to the canton of Geneva. Safe on the other side, they can now thumb their snouts at the French hunters.

The swimming swine show three things. First, they are aware of the danger and can remember the previous year's hunt when family members brought down in a hail of bullets had to be left behind dead or severely wounded. Second, they must be afraid, for this is what drives them to abandon the territory where they have felt so comfortable all summer long. Third, they must be able to remember that they will be safe over in Geneva. Over the extended period of more than four decades, the journey has become a tradition passed from one generation of wild boar to the next: in times of danger, we cross the river to safety. The forefathers of these omnivores discovered this through trial and error in the 1970s. They obviously have a well-developed sense of self-preservation.

We know from the example with the electric fence that animals can become fearful just by remembering something. You know how particular songs, smells, or pictures can dredge up memories of threatening experiences from the depths of your unconscious brain? Well, the same mechanism works with dogs. If you have a canine four-legged family member, you might have had a similar experience to ours. Our little Münsterländer, Maxi, loved life and a change of scene; she just didn't like the vet. When she went there, she had to endure needles and sometimes there was unpleasant tartar removal or the somewhat gross squeezing of her anal glands. Little wonder that every time she went, Maxi stood trembling on the examining table, a little heap of misery as she submitted to each procedure. But not only that. While she was still on her way to the vet, the little dog registered the telltale smells of the neighborhood through the car's ventilation system, and she began to be afraid the moment we turned into the parking lot. A specific film must have been playing through her head that anticipated the unpleasant situation. So let's grant that there is good evidence to show that animals can feel fear. But our little dog's reactions point to something else, as well. Given that there was sometimes a gap of more than a year between visits to the vet, it is clear that dogs, and other animals, can remember things for a very long time (like our goats and the electric fence).

However unpleasant it sounds (and is), most wild animals are like Maxi. As soon as they spot us, they get anxious, especially when we're too close. Apart from fear, it would be interesting to know what other reactions we trigger in them. Can they tell us apart from other animals? Do they have any inkling that we build computers, drive cars, and tower above them intellectually, at least in some respects? Think about it from our point of view. Apart from our pets, no single species of animal has a particular significance for us, none stands out from the others. So, does that mean it is all the same to a deer whether it sees a human, a hawk, or a hedgehog? Theoretically, yes. You can test this out for yourself when you think back to the last time you walked in the forest. Perhaps you remember rare or particularly large or colorful species, but can you remember every bird or describe every fly? You can't, because we take it for granted that our surroundings are full of life, and so we don't bother to notice details about all the creatures crawling and flying around us.

It's difficult to get a clear idea of how other people see the world, because it's almost impossible to get inside their heads. If we can't even do that, how can we hope to see things from the perspective of animals? The easiest way to try to do this is by looking at animals' reactions when we appear. What's important here is whether or not we play an important role in their daily lives. On the one hand, we could have an impact because of the pain or even death we inflict while exploiting or hunting them; on the other hand, we could have an impact because of positive aspects of caring for them, such as providing them with food. The situations I find particularly gratifying are those where we are not involved at all—when we are neither harming nor helping them. In these circumstances, animals usually behave the way they do in representations of paradise: they basically ignore us. A particularly extreme example all the way from Africa was shared on the internet in summer 2015. The online edition of Britain's *Daily Mail* posted photographs from Krüger National Park in South Africa. Lions were standing in the middle of a busy road tearing the flesh off an antelope with cars all around them. What surprised and shocked drivers most was that the predators were completely uninterested in the backdrop to their meal: bushes, stones, people in cars—it was all the same to them.[68]

Less traumatizing experiences of animals ignoring people are the photo safaris you can take in Africa where you can park close to zebra, wild dogs, and antelope. On the Galapagos Islands, on the coast of Antarctica, in the marinas of California, and around Yellowstone—all over the world, you can find places where animals allow people to get very close without getting wary. So why isn't that the case where I live in Central Europe? After all, it has one of the highest concentrations of mammals in the world. There are about 130 roe deer, red deer, and wild boar per square mile of wood. And although theoretically you should be able to see these animals at any time of the day or night, you usually come across them only at night. And you've already learned the reason for this: in Central Europe these animals are hunted wherever they appear.

People are visual animals, and we hunt by sight. And so the goal of potential prey has to be to disappear beyond the limits of this sense. If we hunted by scent, then perhaps over generations animals would have become almost odorless. If we hunted by hearing, they would perhaps have become extremely quiet. Instead, they endeavor to remove themselves from our field of vision. The most important element here is time of day. It's quite simple: because we see almost nothing to absolutely nothing in the dark, our prey evade us by being active at night. We think deer and wild boar are naturally nocturnal, but they're not. They need food at regular intervals around the clock. Instead of spending their days in meadows or along the edges of woodlands, which would be their normal behavior, they spend their days feeding out of sight in underbrush or deep woods. They feel confident enough to leave their hiding places only when darkness falls—the time when people are optically challenged. It is only the very hungry or careless youngsters who venture out earlier and dare to enter areas where hunting blinds loom over the landscape. We euphemistically call these "raised hides," but to deer, they are death traps where their worst enemies perch and deal sudden death with a bang and a puff of smoke. And that is not just my view. It is very clear to colleagues of mine and to hunters that wild animals learn from experience.

This is how a herd of deer experiences the shooting of one of its own. There's a crack and a sudden smell of blood. The shot is often not a clean one, and the wounded animal manages a few yards of panicked flight before its legs buckle and it falls. This sight, coupled with the smell of stress hormones, imprints itself deep into the consciousness of the other members of the herd. And then there's a creaking and crashing from the hide as the hunter climbs down to salvage the downed game, and the intelligent animals make the connection. From then on, before stepping out onto the path, they look suspiciously up at the raised hide to see if someone is sitting there. They could, of course, give the area a wide berth, but hunters usually set up their hides in places where particularly tasty food grows. And if there isn't any, hunters simply sow an appropriate mix of attractive meadow plants. These mixes have names like "Gameland Goulash." Doesn't that sound enticing? And so the evenings become a never-ending game of roulette. If hunger wins, then the deer step out onto the path too early and therefore into the shooters' visual range. If fear wins, the hungry deer don't come out to the feast laid out for them until it's pitch dark, and the hunters go home empty handed.

Foresters in Eifel National Park have observed just how sensitive deer are to their surroundings. A forester who hunts in the area and a forest service employee who works there owned the same make of car. Although the game beat a retreat as soon as the forester's car appeared, they stayed put when the forest service employee drove down the road. But it's not just deer that have mastered the skill of telling the difference between dangerous and harmless people. Our domestic animals also rely on their gut instinct here: deer and company want to avoid hunters, whereas—as we've just seen with Maxi—dogs and cats want to avoid the vet. Hunters are considerably more dangerous than vets, and so it's no wonder that quite a few species of animals can tell what kind of people are out and about. Take the Eurasian jay. Whereas almost all animals consider children to be harmless, jays tend to tolerate adult hikers, as well. That is, until hunters are on the loose. Then jays kick up an enormous racket and warn the animals nearby with their raucous calls. This is why many hunters continue to target these colorful birds, which is unfortunate because jays play an almost indispensable role distributing tree seeds in the woods.

Wild game animals are stressed when people enter their territory. If two-legged intruders are constantly entering the area, the portion of time an animal needs to spend checking to see if it is safe rises from 5 percent to 30 percent of the day.[69] At least that holds true for those people whom animals find it difficult to categorize. Walkers, cyclists, or horseback riders who keep to the trails are easy to evaluate: they make noise and move along clearly demarcated routes. As long as the game can see them, it's clear they are moving directly from A to B, and animals observing them from the safety of their daytime hiding places have nothing to worry about. In contrast, mushroom hunters, mountain bike riders, hunters, and foresters often travel off the beaten track. And because most of these individuals travel alone, there's no lively conversation that wild game can listen in on and use to estimate the route the intruders are taking. All the game can hear is the occasional snapping of a twig under a boot and perhaps a quiet clearing of the throat from time to time—nothing more. Then animals become uneasy, and to be on the safe side, they prefer to move away quickly.

You might argue that things have always been this way. What difference does it make whether it's a wolf pack or a person on the hunt? Well, one big difference is the number of hunters. Contrast one four-legged hunter for every 20 square miles in wolf territory, with about ten thousand two-legged predators tripping up over each other in the same amount of space in Germany today. And the wild game can't tell right away that not everyone is armed. So if there's any doubt, game retreats from each potential attacker and generally forgoes forays into lush meadows in the bright light of day. As you can see, life for game that can be legally hunted is quite nerve wracking. Their situation, where for every potential prey there are any number of potential hunters, is unique in the animal world, because the situation is usually the other way around.

Thus, it's hardly surprising that fear and mistrust are widespread out in the woods and meadows. Let's take a look at which animals have to endure the stress of being hunted. I've already mentioned red deer, roe deer, and wild boar. As far as mammals go, you can add to this list chamois, mouflon sheep, foxes, badgers, hares, martens, and weasels. There are also quite a few species of birds such as partridges, various pigeons, geese, ducks, gulls, snipe, gray herons, cormorants, and corvids. Is it any surprise that we hardly ever get to see any of the birds in this colorful array? Imagine if the situation were reversed and there were between 5,000 and 7,500 lions wandering through each square mile in Central Europe. That roughly corresponds to the numerical dominance of two-legged predators that confronts hunted game in this area. And now let's get back to the animals' point of view. Here—I don't know about you—but my powers of imagination at least have reached their limit. If there was deadly danger lurking in every corner and behind every bush, I wouldn't dare leave my house. If I had to, I would come out only at night when I knew that my persecutors would definitely be sleeping or, if they were up and about, at least they wouldn't be hunting me down.

An animal that has seen a family member collapse covered in blood, or has experienced fear and rising panic deep in its bones, will pass these experiences on, probably over many generations. Researchers have concluded this transfer happens even in the absence of language, for—as reported back in 2009—fear is not only felt in the bones but also expressed in the genes.[70] The Max Planck Institute for Psychiatry in Munich discovered that during traumatic experiences particular chemical markers (methyl groups) get attached to genes. They work like switches and alter the activity of the genes.[71] According to the discoveries researchers made using mice, this means behavior can be changed for life. The research also predicts that, thanks to these altered genes, certain patterns of behavior can be inherited. In other words, our genetic code passes down not only physical characteristics but also, to a certain extent, experiences. And what experience could be more traumatic than the severe wounding or death of your next of kin? It is not a pleasant thought that the majority of the animals living around us are traumatized.

Thankfully, the cohabitation of wild animals and people also has its pleasant side. There is hope that even in Central Europe we can live peaceably with one another, as evidenced by the increasing number of wild animals living in urban areas. The news is going around in animal circles that there's a kind of sanctuary set up in cities. And indeed, built-up areas do count among the so-called no-go areas where hunting is banned. From the animals' point of view, the only difference between cities such as Berlin, Munich, or Hamburg and national parks is that the land has been developed. And what animals do you find in these urban centers? Wild boar in the front garden happily rooting around in the tulip beds ignoring attempts to drive them away (why would they leave?); foxes digging their dens in roadside embankments; raccoons setting up house in garages and attics. The animals are making themselves at home right in the middle of the spaces we've tamed. Whereas to us, asphalt and rows of drab houses symbolize a disconnect with Nature, what animals see is an environment extraordinarily rich in vertical rock faces, where all the mountaintops are bizarrely cube shaped. Urban areas are increasingly revealing themselves to be ecological jewels. Berlin has one of the highest populations of goshawks in the world—about a hundred breeding pairs.[72] The birds nest in the city's parks, using them as a base from which to hunt rabbits and pigeons. I myself have seen a fox near the Brandenburg Gate calmly chowing down on a discarded curry sausage.

Not every city dweller can deal with this proximity to Nature. An older woman explained to me that she was frightened a fox might appear at her patio door. Thoughts of rabies or fox tapeworm flashed through her mind, spoiling the anticipation of what might turn out to be a wonderful wildlife encounter. Wild animals in Central Europe present few dangers. Rabies was eradicated many years ago, and fox tapeworm is quite rare, at least in the wild. I've already mentioned how the chain of infection passes from mouse to fox and the problems with fox feces. If a dog eats an infected mouse (and there are lots of dogs that hunt mice), then it too excretes thousands of eggs when it does its business. Moreover, after it licks its rear clean, it licks its fur, meaning it can distribute tapeworm eggs no bigger than specks of dust throughout the house. A dog that is not regularly dewormed poses more danger than a passing fox.

Perhaps we exaggerate dangers from the wild simply because if we didn't, there wouldn't be much left to fear. Do our archaic instincts simply need something "dangerous" to react to? Granted, it's a bit different with wild boar when they have piglets. A friend of mine who lives in Berlin-Dahlem told me the animals won't leave his yard even when he tries to scare them away by clapping loudly—there's not much more a city dweller can do.

The kite, a large bird of prey, is another European species that seeks out proximity with people and has preferences about whom to hang out with. In earlier times, the birds were hunted and persecuted; however, since they've been protected, they like to spend time around people—that is, they like to spend time around people who own tractors. When the fields are cut in summer, the kites profit from farmers' labors. For the heavy machines don't only cut grass, they also dispatch numerous mice and other small animals into the great beyond. That doesn't sound nice and it isn't good; however, for the kite, this is literally found food. In Hümmel, you can spot the majestic birds as soon as the tractors drive into the fields to begin work. With a wingspan of just over 5 feet, they glide low to the ground, following the machines and keeping an eye out for flattened mice or ground-up fawns.

Martens are a less welcome sight, even though they are really beautiful to look at. They are not hunted in urban areas and trapping, which used to be common in woods and fields, has declined considerably, so martens have mostly lost their fear of people. We once raised an orphaned animal that allowed us to stroke it and made a kind of contented purring sound, just like a cat when it feels good. At first we fed the youngster from a can, but then to prepare it for life in the wild, we also gave it mice for breakfast. It didn't take long for the little chap to become so wild that we could only hold it when we were wearing gloves. Finally, we opened the door to its cage so it could decide for itself when to leave us. It took three nights, and then the cage was empty and we never saw the marten again. But perhaps it's still scampering around somewhere on our property in the dark, although if it were it would be over ten years old by now.

It's questionable whether we did ourselves a favor when we helped the marten. There are two cars parked in front of our forest lodge: a four-wheel-drive vehicle for work in the woods and a car for private use. One day I noticed a piece of rubber hose lying on the ground near the hood of the jeep. I lifted the hood right away and found a nice mess. A marten had done a thorough job of biting through a number of cables and hoses. There was nothing for it but to take the vehicle in to the shop for repair.

But what had induced the animal to go on a rampage in the engine compartment, and why is the marten sometimes gripped by such a destructive mania? By the way, there's no such thing as "the" marten, for there are two species in Central Europe: the pine marten and the beech marten (also known as the stone marten or the white-breasted marten). The pine marten is a shy denizen of the woods that likes to sleep in hollow trees and spends the rest of its time scurrying through the branches up in the canopy. The beech marten, however, is not so attached to trees and feels at home in other places, as well: rocky areas and caves or even houses, which, after all, are just square mountains. These are the places the curious beech marten looks for prey and, while it's doing so, it investigates everything with its sharp little teeth.

Severed cables, trashed hoses, and ripped blanket insulators in the engine compartment, however, are not the result of curiosity but of blind rage. And when these little beasts sense competition, they can certainly lose it. Martens use their scent glands to mark their territory, sending a clear message to every other marten of the same sex: "Occupied. Keep out." Normally, martens respect aromatic boundaries and leave each other alone. Because it's so nice and cozy under the hood of your vehicle, "your" house marten makes regular visits to your car. Sometimes it drops off a few supplies. (One day we found the lower leg of a rabbit on our car battery.) However, these visits don't result in any damage. It's only when you park your car somewhere else overnight that things get serious. In the new location, other martens come on over and investigate the unfamiliar object, leaving scent trails as they rummage through the cavities.

Back on home ground, you've given your marten quite a puzzle. It has to assume that another marten has broken all the rules and entered its favorite hollow uninvited—the absolute nerve! It furiously tries to remove all traces and banish its rivals. Soft hoses are ideally suited for venting rage, and the marten doesn't nibble on them carefully as it does when it's checking things out. In its blind fury, it brutally yanks them out. You can often see from the state of the blanket insulators underneath the hood just how incandescent with rage they are. Sometimes all you find are claw marks; however, when we looked under the hood in our old Opel Vectra, we found the material hanging down in shreds. To achieve this effect, the marten had clearly lain down on its back and thrashed around wildly, ripping out whole chunks of the mat with its sharp claws. The so-called car martens don't necessarily love cars, they simply hate competition. If you park your car in the same place every night, you'll probably be fine.

Meanwhile, there are innumerable insider tips on how to scare the animals away. Small bags of human hair or sanitizing toilet rim blocks hanging in the engine compartment are examples of remedies that will be effective for a few days at most. For a while, we tried sprinkling pepper over the engine. That didn't work in the long term, but a built-in device that delivered a shock when the marten stepped on electrified plates did the trick. We placed the plates where the marten usually got in, and after its first contact, it avoided them. An ultrasonic apparatus equipped with a bright light that reacted to movement was similarly effective. However, the animals become deaf to devices that broadcast ultrasonic sounds all the time, and the constant noise is detrimental to bats and other species, so I wouldn't recommend using them.

And what about our domestic animals? Do they idolize us and stay with us of their own free will? Or is it perhaps fear that keeps them close? If there's a fence involved, the question doesn't even need to be asked. Cows, horses, and, yes, our goats as well are, strictly speaking, prisoners—even if they probably don't think of themselves that way. There's another unpleasant comparison here that has to do with the Stockholm syndrome. The person who coined the term, criminologist and psychiatrist Nils Bejerot, was investigating the relationship between a man who robbed a Swedish bank in 1973 and his victims. The hostages developed feelings for the thirty-two-year-old kidnapper that were similar to the feelings children have for their mothers, whereas they hated the police and the authorities. This paradoxical development has been noted in many similar situations and is considered to be a protective psychological reflex that helps the victims survive their dangerous situations and somewhat lessens the psychological damage they suffer.[73]

If animals are sensitive souls (and I think they are), perhaps they develop similar strategies. When they're held captive, they usually don't trust us right away; they are wary and keep their distance. It takes a while before they start calling out a friendly greeting when they spy us in the distance walking toward the pasture. Does that sound somewhat heartless? Nature didn't intend for goats and horses to spend their whole lives as prisoners behind a fence. Let's not pretend: these animals would hightail it in a heartbeat if they could. So, if they do indeed develop some kind of Stockholm syndrome, it might actually be the best thing for them, because they would then accept their fate and find it tolerable.

When we work out in the pasture, we can tell that our goats and horses enjoy being around us. Of course, the friendly greetings when we appear could also have something to do with feeding time, in which case, we'd only be getting cheered on as bearers of nourishment. It's a bit different with dogs and cats, although not initially. The relationship starts with an involuntary match-up. Then the animals are taken to their new home and forced to spend a few days under house arrest or constrained at the end of a leash when out walking until they get used to their owners. This means the period of acclimatization is not completely of their choosing; however, when dogs and cats finally regain their freedom, they could just take off. But they don't.

More moving are the rare cases when an abandoned pet adopts a person. Completely unforced relationships like this demonstrate that true partnerships are possible. Moreover, these kinds of partnerships happen not only between people and animals, but also between animals of different species. Wolves and ravens are one example, as wolf researcher Elli Radinger explained to me. According to Radinger, ravens enjoy living with wolf packs, and the wolf pups even engage the black birds in play. When large enemies such as grizzly bears approach, the ravens warn their four-legged friends. The wolves repay the debt by allowing their feathered partners to feed alongside them at their kills.

32

HIGH SOCIETY

HAVE YOU READ the novel *Watership Down*? It's a moving story about some rabbits living in the English countryside that have to move out of their old home and find a new one. When they find new territory, they must fight the local rabbits until they finally conquer a corner for themselves. We shelter a family of rabbits in the garden at our forest lodge. Hazel, Emma, Blacky, and Oskar live in a small enclosure. They have weatherproof accommodations and an open space where they are free to run. We get a good view of their social life. There are fights and squabbles, but mostly there are moments of tenderness. The rabbits lick each other's fur or, on warm summer days, they stretch out nestled up against each other in the shade. Of course, there is also a hierarchy, but with only four animals, we can't find out much about it.

Dr. Dietrich von Holst, a professor at the University of Bayreuth, has a completely different set-up. He created a 5½-acre research area for wild rabbits, and he spent twenty years observing their behavior. Population levels were constantly changing. On the one hand, diseases and predators removed up to 80 percent of the sexually mature adults, and on the other, they "bred like rabbits," as the saying goes, so the group grew to about a hundred adults. He noticed that these increases and decreases in population didn't affect all levels of rabbit society equally. Rabbits live according to a strict hierarchy, which is different for each sex. Each rabbit vigorously defends its rank, and for good reason: dominant animals reproduce more successfully. Although the top males and females are more aggressive, overall they suffer less from stress. That sounds logical. After all, rabbits that are constantly being pushed around live in constant fear of the next attack. High-status rabbits experience elevated levels of stress hormones only in short bursts when they are attacking. No wonder Professor von Holst reported that the high-ranking rabbits experienced lower levels of stress overall.

In addition, high-ranking rabbits had especially close social contact with rabbits of the opposite sex, which helped them relax. The average life span of an adult rabbit was two-and-a-half years, and there were distinct differences between rabbits at different levels of the hierarchy. Whereas rabbits on the lowest levels often died just a few weeks after reaching sexual maturity, rabbits of high social standing lived for up to seven years. And that was not just because they got more to eat or were picked off less often by predators. No, the most decisive factor was lower stress levels. A life with less worry and therefore more rest and relaxation meant less risk of intestinal illnesses, which are the leading cause of death in rabbits.[74]

33

GOOD AND EVIL

ANIMALS ARE NOT necessarily better-natured than people, because they can definitely be very aggressive—not only to other species but to each other, as a quick look around our yard confirms. From the yard to the road, there are four beehives, and the busy bees fly out into the surrounding landscape to gather nectar. That's exhausting work, because bees must visit 125,000 flowers to get enough nectar to produce a single ounce of honey.[75] They are not gathering their sweet cargo for me, the beekeeper. The honey they make from it will serve as a source of energy for the bees, which have to shiver their way through the cold of winter. If things don't go according to plan over the summer and the stores of honey they lay up are not sufficient, the search is on for more abundant sources of sweetness. Sometimes, the answer is not to find more colorful flowers: a completely different opportunity might come along in the form of a weak hive in the neighborhood.

Scouts test the targeted hive's ability to defend itself, and if it has been weakened by parasites or pesticides, the call goes out to attack. An intense battle ensues at the hive entrance, but the defenders can't hold off the invaders for long. At some point, the attacking force is just too strong, and the aggressor bees stream past the last dying fighters and advance inside. They pounce on the honeycombs and rip off their waxy covers. In no time at all, they pump their honey stomachs full and fly home, carrying with them the good news for other members of their hive that there's more than enough food stored here. In the hive belonging to the weak colony, there's a loud buzzing from the incessant beating of thousands of wings as the plunderers fly in and out. When there's nothing more to collect, absolute silence descends. Unfortunately, just such a drama played out in my yard, and when I lifted the lid of the vanquished hive, what lay before me was a scene of utter devastation. Crumbs of wax on the hive floor were all that remained of the ripped and shredded combs. A few dead bees were lying around. That was it.

Despite the devastation, the attackers were far from satisfied. They had learned that life is much easier when you attack and plunder your neighbors. If the opportunity arises, they'll do the same again with another hive. As a beekeeper, all I can do is separate the belligerent bees by setting up one of the two warring hives a mile or two away to give the bees an opportunity to calm down. Of course, this isn't an option in the wild, where the game continues until evenly matched hives keep each other in check.

It's not only bees that panic shortly before the onset of winter. Brown bears, for example, can't store food for hibernation, and they have to put on a layer of fat by eating instead. If there's not much food to be had in fall, or when the bears are older and can't gather as much, things get difficult—for people, as well as for the bears. The German filmmaker Andreas Kieling told me the sad story of his colleague Timothy Treadwell. Treadwell saw himself as the bears' friend and took no safety precautions around them. One day, he was watching an old male grizzly in Katmai National Park in Alaska. It had clearly not yet put on enough fat to make it through the winter, probably because it was no longer quick enough to catch the salmon it needed. Experts consider such animals to be particularly dangerous. As usual, Treadwell had neither a weapon nor pepper spray with him. The old bear attacked and killed him. His girlfriend, who was right there watching, was horrified, and she began to scream. This "predator call" (a cry of fear made by prey that triggers a predator's desire to hunt) signaled that there was more food to be had, and so the woman ended up falling victim to the hungry bear as well. The couple was later found buried near their tent. We have a pretty good idea of what happened because we have a recording. Treadwell's original intention had been to film the old bear, and his camera was left running. The cover was still over the lens, but the camera picked up the sounds of the couple's final moments.[76]

Back to animal wars. You can only talk about war, as we use the word, to describe conflicts in species that live in large social groups. In the Central European latitudes, that means bee, wasp, and ant colonies, which mount raids like we do. If, however, an animal attacks another individual on its own, then we talk of a fight, something you can see between many male birds or mammals.

Given that animals attack other animals, can they be considered heartless and evil? Sometimes it seems that way. I have two corner windows in my office, and when I look out of them, I can see an eighty-year-old birch tree that grows in front of the forest lodge. The old tree (birches don't live much past a hundred) has already experienced the ravages of time—or I should say of woodpeckers. About 15 feet up, there is a natural nesting cavity that has been used by a variety of different birds over the years. After the woodpecker, there were nuthatches for a number of years, then one day, starlings. These lightly speckled birds began to successfully rear their young in the cavity. One day I heard a loud screeching, and when I looked out of the window, I saw a magpie that kept flying at the tree. Suddenly it landed on the edge of the cavity and grabbed a little starling. It dropped it from the tree to the ground and began pecking away at it. Instinctively, I got up, left everything in the office, and rushed outside. The magpie flew a few yards and let go of its prize. The young starling was completely distraught, but it did not seem to have suffered any grave injury. I got a ladder and carefully returned it to the nest. As far as I could tell, there were no further attacks, and the baby bird got its start on life along with its siblings.

And yet things probably didn't turn out as they should have, and the fault was mine. What right did I have to get involved with the altercation? Okay, I had felt sorry for the little starling, and I couldn't just watch as it was killed. But from the point of view of the magpie, wasn't it just a piece of meat that was urgently needed to feed the magpie chicks? What if one of them had starved as a direct result of my intervention? In that moment when the magpie pulled the young starling out of the cavity, I thought it was evil. But was it really? And what does that really mean—evil? Can a quality like that depend on your point of view? If so, then from the point of view of the magpie, I was the evil one who had prevented the mother or the father from making a kill. From the point of view of its species, the behavior of the handsome black-and-white bird had been beyond reproach. But I, too, am a typical representative of my species, for most human observers finding themselves in this situation would also have taken pity on the starling.

So what would it look like if a similar incident were to take place between animals of the same species? This isn't unusual in the wild, as we can see with brown bears. In this case, it's the males that can be deadly for the cubs. When mating time approaches, males look for females who are ready to mate. Females with cubs are not in the mood, and so males often just go ahead and take care of this. They kill the cubs, and shortly thereafter the mothers are ready for their next pregnancy—a natural response to the emergency. Because mother bears are well aware of the males' intentions, they try to keep their distance from potential admirers. Alternatively, they mate with as many males as possible so that each thinks he might be the father of the amusing little tots and leaves mother and children in peace. Scientists from the University of Vienna have established that promiscuous behavior is not purely for sexual pleasure but is indeed a defensive strategy employed by the females. After observing bears for twenty years in Scandinavia, they confirmed this behavior happened most often in populations where a particularly large number of cubs were falling victim to attacks by males.[77]

Are these male bears evil? The dictionary defines "evil" as "morally bad, reprehensible." Or to put it more clearly, you could say: for an action to be evil, the perpetrator must intend to violate a moral code to the detriment of others. Neither magpies nor bears do that, for their actions are part of the normal behavioral repertoire for their respective species. However, the behavior of the white rabbits we acquired one day was definitely not normal.

We wanted to switch from the typical common or garden mixes we had on our little farm to purebred animals, and we drove a few villages farther on to take a look at white Viennas. These rabbits have soft, silky fur and adorable blue eyes, and we just had to take a few of them home with us. Back at our forest lodge, they got a spacious run, but the idyll lasted for only a few weeks. One day we came into the pen and saw a wretched little shape hunkered down on the ground. It was a small female, and her ears were slashed so badly that they hung in tatters. We felt so sorry for her, and we thought that there must have been a fierce struggle for dominance. Over the next few days, there were more and more rabbits with slashed ears, and as we watched, we confirmed what we had begun to suspect. One of the females had inflicted these brutal wounds on the others with her razor-sharp front claws. Logically, the ruthless lady was the only one still hopping around with her ears intact.

Now, was this a bad rabbit? I believe it was. This wasn't species-appropriate behavior, and it couldn't be defended on moral grounds either. Whichever way you look at it, the rabbit's intentions were evil. After all, the animal must have acted intentionally and the others had done nothing to encourage her to act the way she did. You could argue that this rabbit might have been traumatized by a dreadful experience in her youth that could have driven her to behave this way. Absolutely. But isn't this almost always the case with human evil-doers as well? Every evil act can be traced back to a point where it can be explained and therefore excused. For simplicity's sake, let's apply the same standards to both animals and people and assume that both have at least a basic level of free will when it comes to deciding how to act. It's not only people who have this freedom of choice; lots of animals do, too.

34

HEY, MR. SANDMAN

A REAL SUMMER FOR me includes swifts. They're like swallows but much bigger and much, much faster. Uttering their piercing calls, they race between tall city buildings, hunting for insects or just for fun. Unlike many other birds, swifts spend almost their whole lives in the air. They are so extremely adapted to life above the earth that their legs have atrophied and their tiny feet are good only for clinging onto things. Of course, they also need to breed, and their nests, which they build in cliffs or in cracks in walls, are constructed so they can fly in and out of them with ease. Apart from the time they spend sitting on nests, the birds satisfy all their other requirements on the wing. Even mating often takes place high up in the air, where swifts, like us, abandon themselves to the moment. And having a male clinging to the female's back does nothing for the birds' flying prowess, so mating pairs often end up spiraling down at breakneck speed, and they have to separate in time to avoid being smashed to smithereens.

But I wanted to introduce you to the swift because of another bodily function: sleep. Most life forms (even trees) need to sleep, and birds land in a sheltered place to do so. Our chickens, for example, take themselves back to the hen house at dusk like good little chickens, tramp up the ramp, and settle on their perch, where they snuggle up side by side. They don't have to worry about falling off, because, as with most birds, their tendons shorten when they sit and their toes curl automatically. This means our chickens can hold on tight without expending any energy. Like all birds, chickens dream. When they do, there's a danger they might move around, just as we do, as their nightly movies play. But if they were to move, they might fall from their perch or, for birds living in the wild, out of their tree. And that's why the muscles used for movement are deactivated the moment a bird nods off, so it can spend the night peacefully with its head tucked underneath its wing.

And what about swifts? They never perch, and they spend not a second longer than they have to on the ground or on the nest. If they want to sleep, they do so while airborne. That is highly risky, of course, because sleeping birds aren't in total control of their actions. And so they spiral upward a mile or more to increase the distance between themselves and the ground. Then they begin to glide downward, tracing a wide circle that slows their descent. Finally, they are free to doze for a few moments. They don't have time for anything more, because they need to be wide awake again before the first rooftops loom dangerously close and their situation becomes precarious.

Is this brief shut-eye sufficient for the birds to get any rest? Definitely, because although sleep allows all species to exclude or reduce outside influences so the brain can run its internal processes undisturbed, sleep is a little different for every species. The different phases of our sleep with their varied depths show that even human sleep is not a uniform affair. Our horses, for example, don't need much in the way of really deep sleep. Often just a few minutes are enough, which they take while lying down on their sides looking as though they've been shot. They're so deep in dreamland that they are indeed dead to the world, and their legs twitch as though they were galloping over an imaginary prairie. Other than that, they stay on their feet and doze away a few hours of each day just like the airborne swifts.

It's obvious that, like us, animals sleep. Even tiny fruit flies need to sleep, and when they're asleep, they twitch their legs just like horses. The really exciting question is how they sleep and what they dream about. Our nightly mental excursions take place during the so-called REM phase of sleep. REM stands for "rapid eye movement." In this phase, our eyes move under our eyelids, and if you wake a person in REM sleep, they can almost always remember what they were dreaming about. Many species of animals experience similar nightly eye movements, and the larger their brain in relationship to their body size, the more they have. Because animals can't talk to us, we have to look for other clues to understand what's going on inside their heads. Researchers at the Massachusetts Institute of Technology in Boston looked at rats. They measured the rats' brainwaves while they were in a maze busily searching for food. Then they compared these readings with instrument displays while the rodents were sleeping. The similarities were so clear that, using the data they had gathered, the researchers could even tell where in the labyrinth the sleeping rats were in their dream.[78]

Experiments with cats in 1967 also yielded evidence for dreaming, although in this case the evidence was indirect. The scientist Michel Jouvet from the University of Lyon stopped cats from relaxing their muscles while they slept. Normally the body, including the human body, shuts off voluntary muscle movement to prevent us from thrashing about wildly in our dreams or walking around our bedrooms with our eyes closed. This shutoff mechanism is only necessary in dream sleep. When the mechanism is deactivated, you can observe what the test subject is experiencing in its dreams. Jouvet observed cats in this state arching their backs, hissing, or running around, all while in a deep sleep. Science accepts this as proof that cats dream.[79]

But what does it look like if we distance ourselves from our branch of the animal kingdom, leave the mammals, and consider insects instead? Could something similar play out inside such tiny heads? Can the relatively small number of cells in a fly's brain also produce pictures during sleep? Today there are indeed indicators that these tiny clumps of cells can do more than we ever gave them credit for. As I just mentioned, fruit flies twitch their legs just before they fall asleep, and their brain is particularly active while they are sleeping—another parallel with mammals. Does this mean that fruit flies dream? Their physical reactions suggest that they do, but so far we can only guess what images might be lighting up inside their little heads. Pictures of mushy fruit, perhaps?[80]

35

ANIMAL ORACLES

I HAVE TO ADMIT, I always used to be a bit skeptical when there was talk of animals having a sixth sense. Okay, so in many species individual senses are more pronounced than ours—but can they really be so much more pronounced that they can pick up almost imperceptible clues to impending natural catastrophes? Today, I believe that this sixth sense is a necessary tool for survival in the wild, a tool that has not been completely lost in the artificial surroundings of our modern world, but which certainly has been buried.

The word "buried" is key, for who, for example, would want to be buried alive in a volcanic eruption? Goats, it seems, are particularly afraid of this eventuality, or so you might think given their ability to detect volcanic activity. A researcher at the Max Planck Institute, Martin Wikelski, tracked this down by tracking the goats. After he fitted a herd of goats on the Sicilian volcano Mount Etna with GPS transmitters, he noticed there were several days when the herd suddenly became restless, as though a dog were harassing them. The animals ran back and forth and tried to hide under bushes or behind trees. Each time, a few hours later, there was a sizable volcanic eruption. The herd gave no noticeable early warning signals for minor eruptions—and, indeed, why should they bother? How were the goats aware of what was about to happen? Unfortunately, researchers don't yet have a satisfactory answer to that question, although they believe it has something to do with the gases seeping up out of the ground just before an eruption.[81]

Animals in German forests are aware of similar dangers. Volcanic activity is an issue everywhere in Central Europe, and you can see signs of it in the Eifel, where I live. Ancient volcanoes tower over the landscape, interspersed with a few young ones, like the one that formed Laach Lake. Young, in this context, means that the last time the volcano erupted was thirteen thousand years ago, and it could do so again at any time. Back then, it spewed 3.8 cubic miles of debris and ash up into the air, shook Stone Age settlements, and formed dust clouds that obscured the sun as far away as Sweden. So it's a danger to be taken seriously, even if there's only a slim chance that we'll experience an eruption in our lifetime.

Where I live, wood ants are the focus for researchers—or I should say for some researchers. Professor Ulrich Schreiber from the University of Duisburg-Essen and his team went to incredible effort to map over three thousand ant hills in the Eifel. They discovered a striking correlation between the placement of the ant hills and faults in the Earth's crust caused by volcanic eruptions and earthquakes. At the intersections of these lines of disturbance, they found concentrations of ant hills. The composition of the gas seeping out of the ground in these places was distinctly different from the composition of the surrounding air. The red wood ants love the mix and prefer to build their homes at these junctions.[82] Now that I'm aware of this, I think of it every time I'm out in the woods and see these attractive formations full of scurrying ants. Scientists don't have the answer to the goat riddle yet, and neither do they know why the ants like these places. The one thing that is clear is that the ants can smell the tiny differences in the concentrations of gases, just as the goats can, and there are numerous reports from all over the world about similar phenomena.

Does this mean animals are inherently more sensitive to their surroundings than people? Naturally, there are many species that perform considerably better in individual sense categories. Eagles see better, and dogs hear and smell better than we do. However, the sum total of our senses is so good that on average we are no different from other species. So why is it that in comparison with animals, we are so unaware of changes in our environment? I believe the answer lies in the way our modern home and work environments overwhelm our senses. Most smells, for example, no longer come from woodlands or meadows, but from exhaust pipes, emissions from the office printer, or the perfumes and deodorants we use on our bodies. Permanent sensory overload as a result of fabricated fragrances masks natural aromas. It's different out in the country only if you spend a lot of time out and about in Nature. For instance, where I live, you can smell a single moped puffing and discharging exhaust from its stinky two-cylinder engine from 50 yards away, and when it rains, the forest air is immediately full of mushroomy smells that forecast a rich harvest in a few days.

It's similar when it comes to keen vision. People who spend a lot of time sitting in front of a computer screen when they're young or surfing with smart phones are more likely to be shortsighted than children who spend most of their time outside. Recent studies show that shortsightedness has increased significantly in younger generations, with rates up to four times higher in people born in the 1960s compared with people born in the 1920s.[83] Are we losing our distance vision? Luckily we can put on glasses; however, the increasing deterioration of our natural acuity of vision seems to me symptomatic of something else.

I believe the senses we are born with are as attuned to natural processes as animal senses are. It is modern life that dulls one sense after another. Even my ears aren't the best these days as a few frequencies have fallen victim to youthful visits to discotheques or to shooting practice. But there is still hope. When something is gone organically, it cannot be repaired; however, our brain can compensate for many things. For me, one beautiful example is the annual migration of cranes. I can hear these birds from a long way off, even through windows well insulated for sound, because I'm always excited to hear these ambassadors of seasonal change. All it takes is a faint suggestion, more like a premonition, and I step outside and see a V of birds flying in the distance.

Migrating cranes are closely tied to the subject of this chapter—the early warning systems of animals. The birds are clues to distant weather patterns, because they like to fly with the wind at their backs. When they fly down from the north in fall, it means bitterly cold north winds are blowing south, which might signal the first snows of the season. In spring, however, the mass arrival of cranes signals the start of the breeding season, meaning that warm southern breezes are drifting north from their winter quarters in Spain, and temperatures will soon be on the rise.

You can even roughly estimate the current temperature by listening. That sounds bizarre, but it's actually quite simple. Grasshoppers and crickets are our little helpers here. These cold-blooded insects begin their concerts when the temperature reaches about 53 degrees Fahrenheit. The higher the thermometer climbs, the faster they chirp. You could argue that a better way to tell the temperature would be by how hot the air feels on your skin. True, but when you're physically active, that's difficult because of the additional heat generated by your body.

Just as you can train your ears, you can also train your eyes. You can correct defective eyesight with glasses, but it is the brain's reaction that is even more important, and you can hone its ability to pick up on particular changes in the surroundings. These days my peripheral vision picks up deer simply by sensing an aberration in the green of the trees. The slightest difference in color in spruce attacked by bark beetles catches my attention before clear differences appear between affected trees and the healthy crowns of neighboring trees.

Other senses of mine have been trained, as well. Whether I can feel a change in the direction of the wind blowing on my face, signaling a change in the weather, or the small droplets of rain that announce light cloud cover (and therefore no serious rain), or minuscule anomalies in the smell of the air that betray the presence of a rotting animal carcass in the distance—together these sensations are pieces of a puzzle that constantly update me on the current state of my surroundings and its dangers without me having to think too much about it.

If you belong to that segment of the population that has a feel for the weather, you too could forecast a storm long before the first clouds appear in a bright blue sky. Scientists don't agree where this sensibility lies—for example, whether it has something to do with the changing conductivity of cell membranes—but, whatever it is, it works. How much more accurately must early peoples have been able to read the woods and the meadows, exposed, as they were, to all those stimuli day in and day out? In my case, I spend only part of the day training my senses like this. Animals train all their lives. No wonder they are so much better at predicting natural disasters.

If animals are so sensitive, what about long-term forecasts? Can animals feel if the coming winter will be a hard one? Some years, you can see squirrels and jays burying particularly large quantities of acorns and beechnuts. I'm sorry to have to tell you that it would be wrong to conclude they are acting with wise foresight to tide themselves over a long and snowy winter. The animals are simply making the most of an overly abundant offering of food the trees have put at their disposal. Beeches and oaks all bloom together every three to five years. This synchronous blooming often happens the spring after a very dry and difficult summer. The bountiful nuts and the squirrels' and jays' diligent nut-gathering happen later that year, which means that their activity attests to what the weather was rather than what it will be, which is a bit disappointing.

So animals are not capable of forecasting the weather in the long term. However, the situation is quite different when we turn our attention to short-term changes. And my favorite species in this respect is the chaffinch. These birds like to live in old deciduous forests, but they also hang out in mixed woodlands. There the male trills a pretty song with a rhythm I remember learning in school that goes something like "chip chip chip chooee chooee cheeoo." But you can only hear this song in good weather. If dark clouds are massing—the first drops of rain might even be falling—then what you hear is "run run run run run" sounded on a single note. I've noticed during my daily forest rounds that the chaffinch changes its song when it's disturbed, but it keeps singing its regular tune when it sees me. It clearly registers the sudden appearance of a person from behind the trees as less of a threat than the sudden disappearance of the sun behind storm clouds.

And what benefit do other chaffinches get when one alert bird notices a change in the weather and warns the others? Couldn't they just all look up into the sky and see the storm front for themselves? Not if they are under the dense leafy canopy of an old beech forest. Here the most they would notice is a slight darkening of the day. The only place to see the approaching menace is through a gap created when one of the giant trees fell and opened up a view of the sky or from way up in the crowns of the trees. Thus the warnings are useful because not every finch is privy to this "insight" from where it is sitting.

36

ANIMALS AGE, TOO

IT'S WIDELY KNOWN that animals get a pain here and an ache there with increasing age, just like we do. But what's going on inside their heads as they become increasingly frail? Are they aware of their diminishing physical capabilities? This question is hard to answer directly using science, but, once again, we can get a good idea by observing their behavior.

Horses seem to get more fearful as the years advance, and with good reason. As I've already mentioned, normally horses do very well dozing standing up. They even have a special knee joint to help them do this. The joint locks in place when the muscles relax to prevent the leg from buckling when the horse is asleep. When the horse dozes, it alternates hind legs, switching its weight to the locked leg, while it rests the tip of its other hoof on the ground. This takes the weight off the forelegs, which both remain straight. A horse can doze for hours at a time like this, but it's not real sleep. Just like us, horses need true deep sleep to stay healthy and fit. And to have that kind of sleep, they must lie down on their side with their legs stretched out on the ground. Only then can they drift off into the land of dreams with heightened brain activity and twitching hooves. Sometimes their lower lip also moves as though they were whinnying in their sleep or trying to eat. Once their slumber is over, they have to get back up again. As they have relatively long legs and may weigh over a thousand pounds, this maneuver requires a great deal of strength. First they heave themselves up with their front legs extended; then they push off the ground with their hind legs and stand up.

Creating the momentum to get up off the ground is almost impossible for old horses, and you can see how the animals get really scared of lying down. Even though they would love to relax completely on their side, for safety's sake they remain standing and content themselves with dozing. That's definitely not good, because energy reserves are drained more quickly in the absence of any real sleep. But the animals clearly know very well that by lying down they are putting themselves in a life-threatening situation. A horse that cannot get up again will soon be dead, because its inner organs will cease to function properly (or a predator will come along). As getting up gradually becomes more difficult, the amount of deep sleep the horse gets diminishes at the same slow rate. We can see this with our two mares. The older one, who is now twenty-three, can be found lying down considerably less often than her companion, who is younger by three years. One day it will get to the point that fear wins—and then Zipy will never dream again.

You can also see changes with old female deer. Apart from the loss of muscle, which makes the does look bonier, there are changes in their behavior. Let me just come right out and say it: they get grumpy and fractious. That's understandable given that they might once have led the herd and been revered as royalty. They are still able to bear young in their old age, but the fawns they bring into the world are weak. The old does' teeth have been worn down completely by years of use so they can no longer chew their food properly, and therefore they are often hungry. This means they're scrawny and their milk production drops, as does its fat content (milk fat is produced in the udder), and so the old does' offspring are starving, as well. It's no wonder these fawns are the ones that most often fall victim to diseases or carnivores. And this, in turn, affects the status of the aging deer, as I explained earlier. Under similar circumstances, you'd probably be crotchety all the time as well, wouldn't you?

One subject I've hardly ever read anything about in aging animals is dementia. Our house pets certainly live much longer than they used to, because their medical care has kept up with ours. Our little Münsterländer is a good example. She always had the best in food, she got her shots, she was taken to the vet for treatment when she had an infection, and while she was there she had tartar removed so she didn't lose her teeth. One day, at the age of twelve, Maxi began to stagger. The problem was quickly diagnosed: she'd had a stroke. That hit us hard. Our little dog, who'd always been so spry, was suddenly nearing the end of her life. But the prescribed pills and injections worked quickly, and Maxi recovered.

Maxi's faculties and senses gradually diminished as she lived out an appropriately long phase of old age. At some point she simply stopped making any noise—barking became a thing of the past, which didn't bother us. Quite the opposite, actually. To match that, her hearing then left her, which was a bit more burdensome, because it meant we could only communicate with her when she could see us. But our dog was still a creature that enjoyed life. In her last year, however, she began to lose it until she no longer knew who we were. She also spent hours turning around in her basket as though she was going to lie down, but she never did. She then ate less and lost a great deal of weight, and cancerous growths appeared. It was with a heavy heart that we asked the vet to put her down. Her successor, the cocker spaniel Barry, went through a similar process at the age of fifteen. Apart from the loss of his mental faculties, he also became incontinent, which cost us a great deal of work and enormous quantities of carpet cleaning foam. You can now find therapies and medications to counteract what is technically referred to as "cognitive dysfunction."

I believe all higher animals can suffer from dementia. Cat lovers report similar behavior in their beloved charges, and researchers have found deposits and changes in the brains of these house pets similar to those found in people suffering from dementia. We even had a goat in our herd with dementia. It lost its ability to orient itself and one day, thanks only to our son's diligent efforts, we found it lying peacefully in a stream in the woods. Observations in the wild are few and far between, because animals suffering from dementia are easy prey for carnivores. They isolate themselves from the community of the herd and in doing so, they signal their vulnerability. An animal that is no longer right in the head is mercilessly eliminated. Of course, the same thing happens with predators, only instead of falling prey to other animals, they starve.

But what does it look like when the end is approaching and the gray cells are still fully functioning? Are these animals aware that the end is near? Not many people, but at least some, can foresee their death. Some are ill and predict the time of their demise almost to the week; others are old and tired and simply don't want to live anymore. For these people, death comes as no surprise. It's the same with some animals. Our very old female goats separate themselves from the herd shortly before their time comes so that they can die in peace. When they go into seclusion, they must know what's in store for them. They search out an isolated patch of pasture or the small shelter the herd doesn't use in summer, at least not during the daytime, and here they lie down and die peacefully.

How do I know this? You can see it from the position of the deceased animal. Our favorite goat, Schwänli, for example, settled down comfortably on her stomach with her legs tucked under her. This is a normal relaxed position for goats when they sleep. In contrast, when an animal dies in distress, it thrashes its legs, disturbing the earth around it. It's on its side, its neck is bent back, and its tongue is often hanging out.

From the animal's position, you can tell that it suffered in its last moments. Not our Schwänli. She had clearly anticipated her death and quietly took leave of this life. Such behavior doesn't only make saying goodbye easier for us; it also has advantages for the herd, at least in the case of wild animals. Weak old animals present a danger, because their slow movements attract predators. By separating themselves from the herd in a timely manner, they make sure that younger members of the herd won't be torn apart as collateral damage when the predator comes for them.

37

ALIEN WORLDS

W E OFTEN FIND Nature idyllic and calming because it looks so peaceful and harmonious. Colorful butter-flies flit through flowery meadows, the white trunks of birches tower over the undergrowth as their leaves wave in the breeze. It really is pure relaxation for us, partly because today's wild holds few dangers for people. That is not the case for the creatures that live there, however, and so they view this idyll from a completely different perspective.

If you look, you will find two distinct differences between butterflies and moths. Butterflies are beautifully colored. The peacock butterfly, for example, sports a large eye on each wing to scare off birds and other hungry creatures. Its body and wings are relatively hairless, so the image the attacker sees is shiny and clear. Moths, in contrast, tend to the monochrome. Gray and brown are their preferred colors, because they spend their days dozing on bark and branches, waiting for dark. During the day, they're sluggish and could easily fall prey to birds, which can distinguish any variation in color with their sharp eyes. Woe to the moth whose wing color doesn't match the bark it's resting on, because that means it's landed on the wrong tree. Moths lacking in tree identification skills don't survive the day, or from the moth's point of view, the night.

In order to survive, moths and butterflies even adapt to our culturally manipulated world. For example, there is the peppered moth. It has a wingspan of 2 inches and black specks on its white wings. That's exactly the color of the birch bark on which it likes to rest. That is to say, the trees in England were white until about 1845. After that, the industrial revolution and the burning of coal released so much soot into the air that a black grimy layer built up on the bark. The moths that used to be so well camouflaged now stood out, and hundreds of thousands of them were eaten by birds, with the exception of a few misfits. These misfits had always been there, and, like black sheep, their wings were dark—a death sentence until then. But now the dark moths turned out to be the winners. They were the ones that survived and were the reason that after a few years most peppered moths were black. It was not until the end of the 1960s, when measures were taken to improve air quality, that the situation changed. The birches were cleaner, which meant they were now white again. And that's why by 1970, most of the peppered moths observed were white again, as well.[84]

At night, things literally look different. Colors are mostly unimportant because insect-eating birds sleep the night away perched on the branches of trees. Other hunters arrive on the scene: bats. They rely less on sight when they hunt and more on ultrasound. They make high-pitched calls and then listen to the echo sent back by objects and potential prey. Visual camouflage doesn't help one little bit, because the flying mammals are "seeing" with their ears. Therefore, the moths must make themselves invisible to hearing. But how do you do that? One possibility is to absorb sound instead of reflecting it. And that's why many moths are covered with a thick furry layer that traps the bats' calls or, to put it more precisely, muddles it up and reflects it back all over the place. Instead of receiving a sharp image of a moth, the bat's brain gets a fuzzy something that might just as easily be a bit of bark.

Pigeons also see completely differently from the way we do. It's true that they are visual animals like us, relying heavily on sight, and they need daylight to see. However, in addition to all the details that are part of our lives, they are clearly aware of other things in the air. They see the polarized pattern of light in the sky—that is to say the geometrical orientation of oscillating light waves—and this polarization is directed north. This means that by day pigeons see a compass everywhere they look. No wonder carrier pigeons orient themselves well over long distances and always find their way home.[85]

Once we include bats' hearing as "seeing," we can widen the range with other species, as well, to understand what they feel and the kind of subjective world they live in. With dogs, for instance, the question becomes how strongly their sense of sight, which is somewhat less acute than ours, is enhanced by their senses of smell and hearing. If it is the sum of the impressions that completes the picture, then we don't know what dogs see if we pass judgment on their eyesight alone. If we did that, we'd have to say that dogs are in dire need of a pair of glasses, because the lenses in their eyes don't do a good job of adapting to different distances, meaning that dogs see objects in focus only when they approach within a range of about 20 feet. If they get closer than about 20 inches, they lose focus again. Dogs use about 160,000 visual nerve fibers to create these images, whereas in our eyes there are about 1.2 million.[86] But even with "visual animals," sight on its own is not enough, and you can test this for yourself right now. If you happen to be sitting somewhere noisy with conversation or street noise in the background, just put your hands over your ears for a moment. That you can hear almost nothing is not the point. The point is that the three-dimensional feeling of your surroundings abruptly changes. Depth is lost. To what extent might the picture dogs have of the world depend on their ears, which are fifteen times more sensitive than ours?

I find it endlessly fascinating when I think that every species of animal may see and feel the world in a completely different way, so you could say there are hundreds of thousands of different worlds out there. And many of these worlds are waiting to be discovered, even in the latitudes where I live. Apart from the species I've introduced already, there are many thousands of others in Central Europe that are unfortunately so minute and unattractive that even their distribution has not yet been systematically researched. Regrettably, nothing is known about their feelings either, because if they don't have a recognized relevance for us, little money is forthcoming for research. And when we don't know what's going on inside these little animals, what their needs are, or how they suffer under commercial forest practices, no one is interested in designating conservation areas for them.

I, for one, would be really interested to know more about weevils, for example. This family includes flightless species that have captured my heart. One representative is a tiny brown nipper, only eight one-hundredths of an inch long, that looks just like a tiny elephant. Its hair grows in stripes along its head and back as though it were a devotee of the retro iro or mohawk haircut. Weevils are adapted for life in the rotting leaves of ancient forests, and these forests are distinguished by one thing above all: they hardly ever change. In Germany, these forests are mostly beech. These trees create a stable community and actively support one another. They exchange sugar and information so efficiently through their interconnected root systems that storms, insects, and even climate change hardly affect them at all. It's an easy place for beetles to live as they munch their way peaceably through the dead leaves on the forest floor. These beetles are relict species. They have survived for generations, and if you find them, you can be sure the deciduous forest is centuries old. Why would they wish to travel anywhere else? What use would they have for wings? They don't need to find a new place to live. Thousands of generations can grow old here undisturbed, and I'm happy to report that the forest I manage is one of the places where one species of these fascinating beetles has been found. I have to say that they grow old by beetle standards, because it takes less than a year for these little animals to enter their dotage.

Without wings, weevils cannot fly away, and between birds and spiders, they have more than enough predators. If you're afraid but you can't run or hide, you have to come up with other strategies, and when they're disturbed, weevils simply play dead. Thanks to their patterned leaf-brown camouflage, they're very difficult to spot. Unfortunately, they're also difficult for visitors to the wood to find unless they have a magnifying glass, because weevils are measured in minuscule fractions of inches. In the absence of ongoing research, we can only speculate what feelings these little chaps might have other than fear. Despite the lack of data, it was important for me to mention them as an example of the many species that are not the focus of our attention and yet still deserve recognition. For the diversity of life that surrounds us truly is astonishing: colorful birds, cuddly mammals, fascinating amphibians, and even useful earthworms. There are interesting animals to be seen everywhere. And that is exactly our Achilles heel. We only admire what we can see, but the majority of creatures in the animal world are so tiny that they are revealed to us only with the help of a magnifying glass, and, for the especially tiny ones, we need a microscope.

What about tardigrades (also known as water bears), for example, of which more than a thousand species have been discovered so far? Eight legs, a cuddly body—they really do look like little bears with too many legs—these eumetazoans (belonging to the scientific subkingdom Eumetazoa) like it to be really moist. The species native to Central Europe mostly range between one one-hundredth of an inch to two one-hundredths of an inch in length, and they prefer to live in moss, which also loves water and stores it particularly well. The tiny "bears" scuttle around here and, depending on the species, eat plant material or hunt even smaller life forms, such as roundworms. But what happens when their homes dry out in the hot summer months?

In the forest I manage, the lush green moss at the bottom of thick beeches is often brown and crispy dry come summer, and the little bears have absolutely no access to water. Then they fall into an extreme form of sleep. Only well-nourished tardigrades survive, and fat plays an important role. If moisture is lost too quickly, death follows; however, if moisture evaporates gradually, the tardigrades adjust, dry out, draw their tiny legs up into their bodies, and reduce their metabolic rate to zero. In this state of suspended animation, they can withstand almost anything: neither searing heat nor bone-chilling cold can touch them. Absolutely no biological activity takes place. They do not dream, because that inner projector requires energy to roll. You could say it's a kind of death, which means there's no aging either. In the general scheme of things, tardigrades are not long lived, but under extreme conditions, they can survive for decades, waiting for rain to reanimate them. When rain comes and saturates both the desiccated moss and the tardigrades, it takes no more than twenty minutes for the tiny creatures to extend their legs and get their internal structures back online. Life as they know it resumes.[87]

38

ARTIFICIAL ENVIRONMENTS

VERY DAY THE Earth is negatively impacted by people, and every day we lose more of the wild as it used to be. We have already cleared, built on, or dug up an unbelievable 80 percent of the Earth's land mass.[88] Animal senses, however, are not configured for concrete and asphalt, but for woods, moors, and intact waterscapes. Artificial light is just one example of the many ways we confuse them.

In Europe, at least half the night sky is affected by light pollution. Even a small town with 30,000 inhabitants contributes to an artificial brightening for 15 miles in every direction. The people who live there have little opportunity to observe a pristine starry sky. And it's not just people that are affected. Many species of animals, especially insects, depend on stars to orient themselves when they travel at night. Moths, for instance, rely on the moon when they want to fly in a straight line. For example, when the moon is at its height and they want to fly west, all they have to do is keep the moon to their left. But little moths can't tell the difference between the moon and a cozy lamp adding a decorative touch to a garden at night. Now, as the tiny winged wanderer glides past the tulips and the roses, it immediately gets turned around. The brightest light at night must be the moon, mustn't it? And so it tries to keep this new moon to its left, but the lamp is unfortunately not 238,900 miles but only a few yards away. If the moth keeps flying in a straight line, the "moon" appears behind it, and it seems to the moth that it must have flown in a circle. And so the insect pilot corrects its course to the left to, as it thinks, continue flying straight ahead. This makes the "moon" appear on the correct side, but what's really happening is that the moth is flying in circles around the light. The spiraling flight takes the moth ever closer to the light until it finally ends up at the center. If the artificial moon is a candle, there's a brief "puff," and the moth's life is snuffed out.

But even without this dramatic finale, the moth is doomed. If it tries to fly a straight course all night and keeps ending up at the light, at some point it will have exhausted its energy reserves. Its intent was to fly to night-blooming plants to fill up on nectar, but the few hours it has left to feed have morphed into an involuntary weight-loss program. And if that wasn't nightmare enough, predators have adapted their behavior to take advantage of the new normal. Spiders spin webs under the lights by our front door, because this is where they hit the jackpot. The moth starts its spiral flight around the light and continues until it inevitably lands on the sticky strands, where it is dispatched by the toxic teeth of the owner.

Roads are particularly challenging for wild animals. First off, asphalt is not in and of itself something negative, because insects and reptiles can warm themselves here until their body temperatures are high enough for them to function properly. Dark surfaces heat up particularly well, which means that, especially in the cooler days of spring, cold-blooded animals (which can generate very little heat themselves) are able to get a jump on the day. That is, as long as a passing car doesn't brutally end their sun worshipping. Roads are attractive in other ways, too. Take deer, for example. Many roadside verges are mown regularly, which means they are a good source of tender grasses and other toothsome greenery. And they are particularly safe places for game animals to hang out, as hunting is forbidden there so as not to endanger drivers. No wonder you can see an impressive number of game animals in these unusual habitats at night. Unfortunately, that, along with the fact that there are so many game animals, is the reason for the large number of traffic accidents. The German insurance industry reports about 250,000 collisions involving wild boar, deer, and other wild animals a year—often with deadly consequences for the animals.[89]

Animals should be capable of learning. They should. But there are two main reasons for the constant supply of crash victims. The first reason is youthful carelessness, which is not restricted to humans. For example, when deer pass their first birthday, they set off to find territory of their own. While old established deer often move no more than a hundred yards in a whole day spent snacking on tender raspberry leaves, young animals are chased away until they chance upon unoccupied space. And, given a density of just over a mile of regional roads alone per square mile of countryside, the yearlings have to cross many bands of asphalt before they find a peaceful unoccupied corner they can call home.

The second reason is love. Roe deer bucks in particular go completely berserk at mating time and have only one thing on their mind: sex. In the heat of the summer months of July and August, hormones run amok, and the male deer are constantly straining to hear a seductive peep. Female deer ready to mate make this sound to draw attention to themselves. Because hunters imitate this call using a blade of grass or a leaf (you hold the blade or leaf tightly between your thumbs, bring your mouth close, and blow), the rutting season is also called "leaf time" in Germany. I admit that even I have tricked a roe buck this way, because I wanted to see if it really worked. And it did: right after the first soft peep, a yearling sprang out of the undergrowth looking around to see where the lady of his dreams might be standing. Because the bucks' senses are in a state of complete confusion, if an amorous adventure beckons, they jump out onto the road without even bothering to look. And that's why in the summer more of the collisions with wild game in Germany involve roe bucks, by day as well as by night.

Does that mean that cities are bad places for wildlife? Not at all. Apart from the restrictions and dangers I've mentioned, there are also great opportunities, above all for species diversity. Outside the city limits, fields and pastures are drowning in seas of liquid manure and being transformed into wastelands, and in the forests clearcut logging machines are sawing down one tree after another and compacting the soil beyond repair, while in the cities, between the rows of houses, new, relatively intact habitats are appearing. No wonder a large number of species from the decimated agrarian deserts have fled to these refugia, including thousands of plants. Scientists estimate that about 50 percent of the native regional and national species of plants in the Northern Hemisphere are to be found in cities. This means our suburbs are now becoming hotspots of diversity. Why am I highlighting the distribution of plants in a book about animals? Well, herbaceous plants, shrubs, and trees are food for animals. They are the first link in the food chain, and therefore they are important indicators for habitat quality. And this means that there are encouraging findings to report for animals, as well. For example, 65 percent of all the species of birds in Poland can be found in Warsaw.

Cities are budding natural areas, comparable to volcanic islands that rise out of the sea in turbulent times, naked and desolate, only to be settled by plants and animals as the years pass. What such young habitats have in common is that they are subject to dramatic changes, so it may take many more decades or even centuries in cities before competing species settle into equilibrium. In Berlin, Munich, and Hamburg (the cities I am most familiar with), you can witness a steady, if slow, transformation. At first, a disproportionate number of non-native species gain a foothold, because they are "set out" in—that is to say introduced into—gardens and parks by the city's inhabitants. It takes many centuries for native varieties to proliferate and reclaim neighborhoods. You can look to the US and Italy to follow their progress. The number of non-native plants in the States decreases from east to west, mirroring the waves of settlement by Europeans, and in Rome their numbers have decreased to just 12.4 percent of the whole. The Eternal City has had more than two thousand years to achieve this balance.[90]

You can see a similar trajectory with animals. Generalists like the fox, which adapt to a wide range of environments, have done relatively well, but, overall, animals appear to have more problems than plants because they need larger territories and they are also threatened by cats, other pets, and traffic. And if a species is particularly successful—pigeons, for example—then we are no longer kindly disposed toward its representatives; in some places, we even begin to mount campaigns against them. I find urban beekeeping to be a particularly positive development. In contrast to the open countryside, there is a good selection of plants in bloom within cities all summer long, which means that the number of hives and the amount of honey they produce is steadily increasing. This shows that there must be enough food for butterflies and bumblebees as well. And so we can conclude that urban areas do not necessarily exclude animals. That being said, we mustn't lose sight of the importance of protecting their original habitats, which is another issue altogether.

39

IN THE SERVICE
OF HUMANITY

OST ANIMALS USED by people lead dismal lives. Count-less pigs and chickens in factory farms are regarded as nothing more than generators of raw material. There's no need to discuss whether these animals voluntarily and happily work for us; we can surely agree that they don't. However, there are touching examples of human-animal partnerships that are a joy to behold. I often see these partnerships in action in the forest I manage. I'm talking about the log hauler and horse teams that remove felled trees. These days, most trees are felled using mechanical harvesters. These hugely heavy machines are detrimental to the forest because they crush the delicate soil as far as 6 feet below the surface. Therefore, in my own community forest, we use forest workers to cut down trees. Felled trees then need to be dragged out along logging roads. And in Hümmel, removal is done using heavy horses, just as it was hundreds of years ago.

Do these horses enjoy their work? Don't they find it tedious to spend the whole day dragging heavy loads until the sweat runs down their flanks? First, let's talk about the load. To reduce the weight, forest workers cut the trunks, which can be up to 100 feet long, into 16-foot sections. Not only are these lighter, but they are also easier to maneuver around standing trees. And now it's the log haulers' turn. I've never met one that didn't love his or her animals. Their horses are their colleagues, and they don't want to overwork them. Because there are no holidays or weekends when you're looking after horses, the animals are more like valued family members. When they're out in the woods, the log haulers are careful that nothing happens to their charges, and it is the horses themselves that are eager to work just that little bit harder. You can see how much they enjoy their work when they are forced to take a break.

Log haulers usually have a second horse ready to take over so they can fill a reasonable daily quota of logs. In the first half of the work day, at least, the "resting" horse paws impatiently at the ground. It would much rather be back at work than taking a breather. Even when the horses are working, it would be easy for them to balk, because the log haulers usually guide them with a slack rein. The leather lines would be far too weak to stop these giants, which weigh in at a few tons, or to drag them in a particular direction. The reins serve mainly to keep the log haulers in contact with their horses and to pass small signals forward. The rest is accomplished by means of an incomprehensible gibberish, babbling sounds like "Jojo, hejhe, brr." The horses listen and know exactly whether they are supposed to go forward, backward, or sideways, and whether they should pull with all their might or advance carefully. Similar human-animal partnerships are shepherds and their dogs, which also follow verbal commands. This is another example of animals taking pleasure in their work, as you can clearly see if you watch sheepdogs racing around a flock of sheep to round them up.

On the subject of "domestic animals," there are two completely different points of view. One is that we have molded our fellow creatures through breeding to make them perfectly suited to meeting our needs. Wild has become tame, slim has become fat, large has become small—no matter what our desires, animals can meet them. Some species have been formed into bizarre caricatures of their former selves. However, you could also look at this process from another perspective—the perspective of the animals. They have succeeded in changing themselves so that they know exactly how to push our emotional buttons. And this brings us back to Crusty, the French bulldog. This little snub-nosed dog exuded a natural charm—you just had to stroke him. So who was manipulating whom? Food and water were provided. If there was a boo-boo, there was a trip to the vet. In winter, there was always a comfortable spot by the fire. The little chap led a really agreeable life. If he had still been roaming around outside like his ancestors the wolves, there's no way his life would have looked like this.

The example of lactose tolerance also shows how far we have adapted to life with our four-legged companions. Normally only infants tolerate milk, because they are the only reason mothers produce the white liquid. The ability to digest milk, or more specifically lactose, is gradually lost as infants adapt to solid food. Or I should say it used to be lost. When people began to keep domestic animals, it became possible for adults to consume milk and cheese from their cows and goats. Because milk is a valuable food, communities where a genetic mutation made it possible for their members to digest lactose had a better chance of survival. This process can be traced back about eight thousand years and is still ongoing, meaning that in Central Europe only 90 percent and in Asia only 10 percent of the population has this ability. There's no research yet to see how people might have adapted to live with dogs. Depending which scientist you ask, this relationship has been going on for as long as forty thousand years.[91]

40

COMMUNICATION

A S I HAVE already mentioned, in the end we will never know if fear, grief, joy, or happiness feel the same to animals as they do to us. We cannot even say for certain if one person feels the same as another, as you have perhaps discovered for yourself when it comes to pain. You can use stinging nettles to test whether some people are more sensitive than others to the same experience: some scream out loud, while others barely feel anything. And yet, we can talk to one another, so we can get at least some idea of what other people are feeling. We can't do that with animals.

Really? Studies of ravens tell a very different story, as we have already seen with naming. Greeting new arrivals using variously pitched calls—some higher, some lower—gives immediate feedback on the esteem in which they are held. There is no better way to express emotions than this. But communication isn't all about sounds. Even with people, a considerable portion is nonverbal—transmitted by means of facial expressions and gestures. Depending on which studies you want to believe, the verbal content of a conversation might convey as little as 7 percent of its meaning.[92]

And what about animals? Like us, ravens don't rely on sounds alone. Researchers working with Simone Pika at the Max Planck Institute for Ornithology in Seewiesen discovered that these intelligent birds use their beaks in much the same way we use our hands. Whereas we point to something with a finger or raise a hand and wave in order to attract someone else's attention to an object or to ourselves, ravens hold up objects with their beak. They also use their beak to point to something or to attract the attention of a member of the opposite sex. With an extensive vocal "vocabulary" and a number of physical movements that they can adjust to suit the situation, they have an enormously detailed ability to express themselves.[93] They need this because ravens spend almost all their lives together, and they need a way to thoroughly check each other out.

This discovery, however, opens just one small window onto the emotional life of these black birds, which still have many surprises to offer us. There was another such avian "interpreter" at our forest lodge. Our children had been given a pair of budgerigars as a gift, and Anton, the male, knew how to draw attention to himself. When he wanted to be fed, he picked up his bowl and dropped it. He had plenty of other toys in his cage, so clearly this gesture was a goal-oriented message that read: "Please fill this bowl!"

But let's leave gestures and get back to speech. Not only can dogs bark, but they can also make a variety of sounds that they use in a general way to express themselves—or perhaps dogs differentiate more precisely between the sounds, while all we can decode is the general message. We got the impression this might be the case with our Münsterländer, Maxi. As the years progressed, we learned whether she wanted to tell us that she was hungry, or bored, or that her water bowl was empty. And even our horses turned out to be capable of relatively nuanced vocalizations. I was especially surprised by equine research coming out of Switzerland. Most horse owners know that horses communicate a lot using body language. Even though research into equine nonverbal communication is more advanced than corvid research, researchers at ETH Zurich were surprised to discover that there was more meaning in even seemingly primitive calls than previously suspected. They discovered that whinnies contain two basic frequencies and can transmit complex information. The first of the two basic frequencies indicates whether the whinny is communicating a positive or a negative emotion. The second frequency indicates the strength of the emotion.[94] On the relevant page of the ETH website, you can listen to an example of a whinny for both situations.[95]

After listening to the recordings, I knew right away that our horses clearly express a positive emotion when they whinny as they see us coming. True, there's usually food involved, but I'm not concerned about that. What interests me is that I can now say with absolute confidence that the horses are happy to see us, something I used to have to assume. And after reading the research results, I listened more closely to find out whether their calls varied, and whether they were happier to see us at some times than at others. Now I know. Of course, there are times when they are happier, just like people. Aside from this study, I am convinced that there are also whinnies that express affection. When our older mare, Zipy, cuddles with us, she makes soft, high-pitched sounds through closed lips. When she does that, we know she's feeling good and is happy to be with us. In other words, she's communicating her emotions to us "verbally." I think horses are a good example of how little we know about communication among animals. People have kept horses for thousands of years. Therefore in theory they should have been much more thoroughly researched than wild animals. That research, and recent research at that, can surprise us so much makes me more careful about passing judgment on the skills other species possess.

The next step in communication would be if we not only deciphered the language animals use to talk to each other but could also talk to them. Then it would be possible to ask them directly about a wide range of emotions, and we could spare ourselves boring scientific studies. And something like this exists already. There is a female gorilla called Koko who has very moving things to say. Yes, I mean "to say," and she does so using sign language. Penny Patterson trained the young ape as part of her doctoral studies at Stanford University in California. Over time, Koko has learned more than one thousand signs, and she can understand more than two thousand words in English. Thanks to her proficiency in sign language, she can let scientists know what she is thinking, and for the first time it is possible to hold an extended conversation with an animal. Other apes have been trained with similar results, indicating that Koko is not an exception.[96] The female gorilla makes regular media appearances and there are often touching scenes. One time she was given a stuffed zebra as a gift. When she was asked what it was, she replied with the signs for "white" and "tiger." And when she was asked where animals go when they die, she signed "comfortable hole."[97] Koko has given so many intelligent answers—and has combined concepts she has mastered with new ones—that it really does make sense to call her an ape with a gift for language.

However, there are also strident critics of the Gorilla Foundation, the organization dedicated to these great apes whose most important project is exploring Koko's world. An external review of the results isn't possible as the project publishes very little. Moreover, the conversations with Koko are not conducted according to scientific protocol. For instance, the gorilla often makes mistakes when she answers questions, which the researchers put down to her playful nature.[98] Unfortunately, I cannot tell you what is true and what isn't from what has been made public, but a gut feeling tells me that at the very least we have severely underestimated the capacity of our fellow creatures to communicate. And the big question for me is not whether Koko can really speak or whether only some of her answers make sense, because communication between people and animals will always be very one sided. People try to teach other species human language. The species is then thought to be particularly intelligent when its members understand a lot of concepts or commands, and perhaps can even utter a few intelligible words. People are thrilled when budgerigars, ravens, or apes like Koko can answer a question in our language.

If we really are the most intelligent species on Earth—and I believe we are—why didn't science approach this from the other direction a long time ago? Why are years spent painstakingly teaching lab animals sign language if modern researchers believe that their capacity to learn is less than ours? Wouldn't it be much easier if we finally began to learn the language of animals? We have many more opportunities now than we had a few years ago, when it would not have been possible to produce sounds at, say, horse level because we lacked the ability to whinny at two different frequencies. Today, a computer could do a reasonable job of translating what we wanted to say into the appropriate animal vocabulary. Unfortunately, I don't know of any serious attempts to do this. There are certainly people who can imitate animal voices, for example the calls of different species of birds. However, people who can imitate a blackbird or a chickadee can do no more than pipe "this spot's already taken" in bird language. That's all that the beautiful song trilled by the males sitting up there in the treetops means. What sounds so delightful to our ears serves within the species to scare off competitors. That would be like having a parrot that kept saying, "Go away." Unfortunately, that's as far as we've got with our ability to communicate with our fellow creatures.

41

WHERE IS THE SOUL?

SO NOW IT'S time to dig deep: do animals have a soul in the sense of an entity with no physical form? It's a really tricky question and one that I would first like to explore in people because that's a bit easier. So what do we mean by soul? The dictionary, interestingly enough, has a number of different definitions, which shows that there is no common understanding of what the word "soul" means. One definition suggests the soul is "the principle of life, feeling, thought, and action in humans." Another suggests it is the spiritual part of people, which, according to religious beliefs, lives on after death.[99] And because no one can inspect the latter, I'm going to set my sights on the first definition.

Feelings, thoughts, and actions—there you have all you need to describe the essence of an animal, right? Animals definitely act and we no longer deny them feelings, so that just leaves thoughts. According to the dictionary definition (which only applies to people, of course) the capacity to think is a prerequisite for a soul. Okay, let's search for this capacity. That's not easy, because there are also a lot of different descriptions for thought, many of which are highly complicated and yet still don't adequately capture the concept. The Technical University of Dresden offered its students this explanation, among others: "Thinking = a mental process, in which symbolic or pictorial representations of objects, events, or actions are generated, transformed, or combined." A much simpler explanation mentioned in the same context puts it more succinctly: "Thinking is problem solving."[100] According to this definition, thinking is part of the skill set at least of those animals whose behavior makes sense to us: ravens that greet other ravens by name, rats that reflect on their actions and regret what they have done, roosters that lie to their hens, and magpies that risk a bit on the side. Would anyone deny that problem-solving processes are going on in their heads?

And with that I would like to come back to the second, religious definition of a soul, after all. Even if I have to tread carefully and I don't feel confident here, even if belief and logic are to a certain extent mutually exclusive, I would like to argue for the existence of an animal soul in the religious sense of the word. Assuming you don't believe in physical resurrection, a soul is a prerequisite for life after death. And if people have souls in this sense, then animals must have them as well. Why? Because there is the question of how long people have been going to heaven. For two thousand years? For four thousand years? Or for as long as there have been people? That would be about 200,000 years. But where is the break with earlier life forms, with our predecessors? The break didn't happen abruptly. It happened stealthily. Small changes spread out over generations in the course of evolution. Which individuals are then to be identified as no longer being people with souls? Some female ancestor who lived 200,023 years ago? Or a man armed with a flint weapon who lived before the year 200,197? No, there is no sharp demarcation, and so you can trace the lineage further and further back, past our primitive forefathers, primates, the first mammals, dinosaurs, fish, plants, bacteria. If there is no specific point in time, X, to which the genesis of a creature of the species *Homo sapiens* can be fixed, then there is also no specific point in time when the soul appeared. And if there is a higher form of justice in the religious sense, then when it comes to the question of eternal life, there can't very well be a sharp line drawn between two generations, where the older one is denied entry and the younger one is admitted. Isn't it a beautiful vision that up in heaven there will be a throng of animals of different species living among countless people?

Quite apart from that, I personally do not believe in life after death. I envy anyone who does, but my powers of imagination cannot carry me that far. Therefore, I am content with the first description of the soul, which I am happy to attribute to all animals. I simply find it a fine thought that other species are not mere machines in which everything is directed by predetermined mechanisms and actions happen when a button is pushed, which is to say when a hormone is released.

Squirrels, deer, or wild boar with souls. That's the thought that makes life special and warms my heart when I have the opportunity to watch animals like these in the wild.

EPILOGUE

WHEN I LOOK at animals, I like to make analogies to people, because I cannot imagine that animals feel so very
differently from us, and there's a good chance I'm right.
The idea that there was an abrupt break in the course of evolution and that at some point everything was reinvented is an
idea whose time is past. The only major point of contention
today is whether animals can think; that's what we do best,
after all.

However, what is so important for us could be less important for our fellow creatures—otherwise they would have
developed like we did. Is profound thought something that is
absolutely necessary? It's certainly not necessary for a satis-
fying, peaceful life. When we're relaxing on vacation, what's
going through our mind is: "I feel great and I don't have to
think about anything at all." We can experience joy and peace
without giving anything much thought, and that is the crux

of the matter: emotions have no need for intelligence. As I have stressed, emotions steer instinctive programming and therefore are vital for all species, and therefore all species experience them to a greater or lesser degree. Whether an animal reflects on these emotions, prolongs them through reflection, or relives them later is less important. Of course, it's nice that we can do exactly this, and perhaps by doing so we get to experience these moments in our lives more intensely. Score one for us. Admittedly, that works for less enjoyable moments as well, so score one for the animals, which makes us even.

Why is there still so much resistance to the idea that our fellow creatures have the capacity to feel joy and to suffer? This resistance comes from some scientists, but above all from politicians who answer to farmers. Mostly they are protecting the cheap methods used by factory farms to house and handle animals, such as castrating piglets without anesthetic, as I mentioned earlier. And then there's hunting, which claims the lives of hundreds of thousands of large mammals and many birds every year, and which in its current form is simply no longer appropriate.

When all the arguments have been made and it's clear we're getting to the point where we must grant animals way more skills than we usually do, the knockout punch is delivered: the charge of anthropomorphism. People who compare animals to humans, so the argument goes, are unscientific. They are wishful, maybe even mystical, thinkers. In the heat of the fray, an essential truth we all learned in school is overlooked: a human being is, from a purely biological perspective, an animal and therefore not so very different from other species. It follows that a comparison between people and animals is not too much of a stretch, especially since we can only relate to and empathize with things we understand. And so it makes sense to take a closer look at those animals in which we can detect emotions and mental processes similar to our own. This comparison is easier with feelings such as hunger and thirst, whereas comparing human and animal experiences of joy, grief, or compassion makes some people's hair stand on end.

The goal is not to anthropomorphize animals but to help us understand them better. Most importantly, these comparisons serve to point out that animals are not dimwitted creatures clearly stuck a level below us on the evolutionary scale, creatures that experience only pale imitations of our rich range of sensations for pain and other such feelings. No. People who understand that deer, wild boar, and ravens lead their own lives, perfect in their own way, and have a lot of fun while they're at it, might even respect animals as insignificant as the tiny weevils that rummage around contented and happy in the leaf litter of ancient forests.

One reason there is still doubt about the emotional world of animals might be that many emotions and mental processes are not yet clearly defined even in people. In this connection, I'll remind you of happiness, gratitude, or even just thinking—all terms that have been difficult to describe. How can we understand something in animals that we can't even clearly grasp in ourselves? Pure science, which today is defined by its demand for objectivity, might not help us advance, because it requires us to set our emotions aside. However, because people are mostly driven by their emotions (see the chapter "Instinct—A Second-Rate Emotion?"), we possess the appropriate antennae to recognize stirrings of emotion in others. And why should these antennae fail just because the other in question is an animal and not a person?

We evolved in a world full of other species, and we had to survive despite them and with them. It was surely just as important to be able to read the intentions of wolves, bears, or wild horses as it was to read the faces of strangers. Certainly, our senses can sometimes deceive us and we can read too much into the behavior of dogs and cats. But I'm convinced that in the majority of cases our intuitions are correct. Current scientific discoveries come as no real surprise to animal lovers. All they do is to give us more confidence to trust our own feelings where animals are concerned.

When people reject acknowledging too much in the way of emotions in animals, I have the vague feeling that there's a bit of fear that human beings could lose their special status. Even worse, it would become much more difficult to exploit animals. Every meal eaten or leather jacket worn would trigger moral considerations that would spoil their enjoyment. When you think how sensitive pigs are, how they teach their young and help them deliver their own children later in life, how they answer to their names and pass the mirror test, the thought of the annual slaughter of 250 million of these animals across the European Union alone is chilling.[101]

And it doesn't stop with animals. As science has discovered and you might well already have read, we must now acknowledge that trees and other plants have feelings and even a capacity to remember. How, pray tell me, are we supposed to feed ourselves in a morally acceptable manner if we are now justified in feeling sorry for plants, as well? Like many species, we cannot photosynthesize to create our own food, so we have to eat living entities to survive. The choices we make are very personal. They might depend on where we live or the culture in which we were brought up. Ultimately, though, each of us has to decide what we will eat. My hope is that what you have learned in this book will help you make informed decisions for the future.

From my personal perspective, I am suggesting that we infuse our dealings with the living beings with which we share our world with a little more respect, as we once used to do, whether those beings are animals or plants. That doesn't mean completely doing without them, but it does mean a certain reduction in our level of comfort and in the amount of biological goods we consume. As a reward, if we then have happier horses, goats, chickens, and pigs, if we can then observe contented deer, martens, or ravens, if one day we can listen in when the ravens call their names, then a hormone will be released into our central nervous system that will spread a feeling against which we have no defense—happiness!

ACKNOWLEDGMENTS

I OWE MY WIFE, Miriam, a huge thank you. Once again, she worked through my unfinished manuscript many times, reviewing the thoughts I had committed to paper with a critical eye. My children, Carina and Tobias, helped jog my memory as I brooded yet again in front of a blank screen completely unable to come up with a fitting anecdote no matter how I tried, even though there were so many to choose from— thank you both, my dears! The team at Ludwig Verlag, my publisher, came up with the idea for a book centered on animals (ah yes, there were so many ideas swimming around in my head that I could have written three books). Thank you!

Angelika Lieke gave the text one final polish, improving its readability by pointing out repetitions, illogical sentences, and impediments to understanding. I don't want to forget my agent, Lars Schultze-Kossack, who made the connection with the publisher and gave me constant encouragement when I was harboring sincere doubts about whether this could ever turn into a book (as he did with my book about the hidden life of trees, when I was equally unsure of myself). And last but not least, I want to thank Maxi, Schwänli, Vito, Zipy, Bridgi, and all the other four-legged and two-winged helpers that have allowed me to be a part of their rich lives—in the end, they were the ones that had all the stories to tell, which, dear reader, I then got to translate for you.

NOTES

1   Max Planck Institute for Human Cognitive and Brain Sciences, "Unconscious Decisions in the Brain," April 14, 2008, www.mpg.de/research/unconscious-decisions-in-the-brain.

2   McGill Newsroom, "Squirrels Show Softer Side by Adopting Orphans," news release, June 1, 2010, www.mcgill.ca/newsroom/channels/news/squirrels-show-softer-side-adopting-orphans-163790.

3   *The Guardian,* "French Bulldog Called Baby Adopts Six Wild Boar Piglets," February 15, 2012, www.theguardian.com/world/2012/feb/15/french-bulldog-wild-boar-piglets.

4   *Toronto Star,* "Yeti the Farm Dog Nurses 14 Piglets in Cuba," September 3, 2011, www.thestar.com/news/world/2011/09/03/yeti_the_farm_dog_nurses_14_piglets_in_cuba.html.

5   Amy DeMelia, "The Tale of Cassie and Moses," *Sun Chronicle,* September 5, 2011, www.thesunchronicle.com/news/the-tale-of-cassie-and-moses/article_e9d792d1-c55a-51cf-9739-9593d39a8ba2.html.

6   Antje Joel, "Mit diesem Delfin stimmt etwas nicht" (Something's wrong with this dolphin), *Die Welt,* December 26, 2011, www.welt.de/wissenschaft/umwelt/article13782386/Mit-diesem-Delfin-stimmt-etwas-nicht.html.

7   C. George Boeree, "The Emotional Nervous System," Shippensburg
    University Webspace, 2009, http://webspace.ship.edu/cgboer/
    limbicsystem.html.

8   Victoria Braithwaite, *Do Fish Feel Pain?* (Oxford: Oxford University
    Press, 2010).

9   Justin S. Feinstein, et al., "The Human Amygdala and the Induction
    and Experience of Fear," *Current Biology* 21, no. 1 (2011): 34–38. doi:
    10.1016/j.cub.2010.11.042.

10 Manuel Portavella, Blas Torres, and Cosme Salas, "Avoidance Response
    in Goldfish: Emotional and Temporal Involvement of Medial and Lateral
    Telencephalic Pallium," *The Journal of Neuroscience* 24, no. 9 (2004):
    2335–42. doi: 10.1523/JNEUROSCI.4930-03.2004.

11 Hubertus Breuer, "Die Welt aus der Sicht einer Fliege" (A fly's eye view
    of the world), *Süddeutsche Zeitung*, May 19, 2010, http://www.
    suedeutsche.de/panorama/forschung-die-welt-aus-sicht-einer-fliege-
    1.908384.

12 Forschungsverbund Berlin e.V. (FVB), "Do Fish Feel Pain? Not as Humans
    Do, Study Suggests," news release, *ScienceDaily,* August 8, 2013,
    www.sciencedaily.com/releases/2013/08/130808123719.htm;
    Spiegel Online, "Fische kennen keinen Schmerz wie wir" (Fish don't
    feel pain like we do), September 9, 2013, http://www.spiegel.de/
    wissenschaft/natur/angelprofessor-robert-arlinghaus-ueber-den-
    schmerz-der-fische-a-920546.html.

13 Marco Evers, "Leiser Tod im Topf" (A quiet death in the pot), *Der Spiegel,*
    52 (2015): 120.

14 Tamar Stelling, "Do Lobsters and Other Invertebrates Feel Pain? New
    Research Has Some Answers," *The Washington Post,* March 10, 2014,
    https://www.washingtonpost.com/national/health-science/do-lobsters-
    and-other-invertebrates-feel-pain-new-research-has-some-answers/
    2014/03/07/f026ea9e-9e59-11e3-b8d8-94577ff66b28_story.html.

15 Jennifer Dugas-Ford, Joanna J. Rowell, and Clifton W. Ragsdale, "Cell-
    Type Homologies and the Origins of the Neocortex," *PNAS* 109, no. 42
    (2012): 16974–79. doi: 10.1073/pnas.1204773109.

16 Chris R. Reid, et al., "Slime Mold Uses an Externalized Spatial 'Memory' to Navigate Complex Environments," *PNAS* 109, no. 43 (2012): 17490–94. doi: 10.1073/pnas.1215037109.

17 American Association for the Advancement of Science, "Slime Design Mimics Tokyo's Rail System: Efficient Methods of a Slime Mold Could Inform Human Engineers," news release, *ScienceDaily,* January 22, 2010, https://www.sciencedaily.com/releases/2010/01/100121141051.htm.

18 Washington State Recreation and Conservation Office, Washington Invasive Species Council, "Feral Swine," http://www.invasivespecies.wa.gov/priorities/feral_swine.shtml.

19 Kirstin Lauterbach, et al., "Do All Male Wild Boar Yearlings *Sus Scrofa* L. Leave Home?" Sixth International Symposium on Wild Boar (*Sus scrofa*) and Sub-Order Suiformes, Kykkos, Cyprus, October 2006. doi: 10.13140/2.1.5146.4962.

20 Statista, "Jahresstrecken von Schwarzwild (Wildschweine) in Deutschland von 1997/98 bis 2014/15" (Annual numbers of wild boar in Germany from 1997/98 to 2014/15), http://de.statista.com/statistik/daten/studie/157728/umfrage/jahresstrecken-von-schwarzwild-in-deutschland-seit-1997-98/.

21 Elke Bodderas, "Schweine sprechen ihre eigene Sprache. Und bellen" (Pigs speak their own language. And bark), *Welt,* January 15, 2012, http://www.welt.de/wissenschaft/article13813590/Schweine-sprechen-ihre-eigene-Sprache-Und-bellen.html.

22 Solon Kelleher, "Tangled Whale Gives His Rescuers the Best Thank You Ever," *the dodo* (blog), July 8, 2015, www.thedodo.com/tangled-whale-says-thank-you-1238266959.html.

23 Katy Sewall, "The Girl Who Gets Gifts from Birds," *BBC News*, February 25, 2015, http://www.bbc.com/news/magazine-31604026.

24 Anders Pape Møller, "Deceptive Use of Alarm Calls by Male Swallows, *Hirundo rustica:* A New Paternity Guard," *Behavioral Ecology* 1, no. 1 (1990): 1–6. doi: 10.1093/beheco/1.1.1.

25 Michael Becker, "Elstern" (Magpies), March 1999, http://www.ijon.de/
    elster/verhalt.html.

26 Rebecca Grambo, *The Nature of Foxes* (Vancouver: Greystone, 1995), 30.

27 Michael A. Steele, et al., "Cache Protection Strategies of a Scatter-
    Hoarding Rodent: Do Tree Squirrels Engage in Behavioural Decep-
    tion?" *Animal Behaviour* 75, no. 2 (2008): 705–14. doi: 10.1016/j.
    anbehav.2007.07.026.

28 Rachael C. Shaw and Nicola S. Clayton, "Careful Cachers and Prying Pil-
    ferers: Eurasian Jays *(Garrulus glandarius)* Limit Auditory Information
    Available to Competitors," *Proceedings of the Royal Society B* 280, no.
    1752 (2013). doi: 10.1098/rspb.2012.2238.

29 Max Planck Institute for Ornithology, "Competition Favors Shy Tits,"
    March 10, 2016, http://www.orn.mpg.de/3685340/news_publication_
    10363050.

30 Christopher Turbill, et al., "Regulation of Heart Rate and Rumen Tem-
    perature in Red Deer: Effects of Season and Food Intake," *Journal of
    Experimental Biology* 214, no. 6 (2011): 963–70. doi: 10.1242/jeb.052282.

31 University of Veterinary Medicine, Vienna, "Personality Differences: In
    Lean Times Red Deer with Dominant Personalities Pay a High Price,"
    news release, September 18, 2013, www.vetmeduni.ac.at/en/infoservice/
    presseinformation/press-releases-2013/press-release-09-18-2013-
    personality-differences-in-lean-times-red-deer-with-dominant-
    personalities-pay-a-high-price.

32 Freie Universität Berlin, "When Bees Can't Find Their Way Home,"
    news release, March 20, 2014, http://www.fu-berlin.de/en/presse/
    informationen/fup/2014/fup_14_092-bienenorientierung-pestizide-
    publikation-menzel/index.html.

33 Stefan Klein, "Die Biene weiss, wer sie ist" (The bee knows who it is),
    Zeit Magazin, February 25, 2015, http://www.zeit.de/zeit-magazin/
    2015/02/bienen-forschung-randolf-menzel; Randolf Menzel, et al., "A
    Common Frame of Reference for Learned and Communicated Vectors in
    Honeybee Navigation," *Current Biology* 21, no. 8 (2011): 645–50. doi:
    10.1016/j.cub.2011.02.039.

34 Alan Bellows, "Clever Hans, the Math Horse," *Damn Interesting*, article 170, last updated November 30, 2015, https://www.damninteresting. com/clever-hans-the-math-horse/.

35 Andreas Lebert and Claudia Wüstenhagen, "In Gedanken bei den Vögeln" (In the minds of birds), *Zeit Online*, July 20, 2015, http://www.zeit. de/zeit-wissen/2015/04/hirnforschung-tauben-onur-guentuerkuen; Lorenzo von Fersen and Juan Delius, "Long-Term Retention of Many Visual Patterns by Pigeons," *Ethology* 82, no. 2 (1989): 141–55. doi: 10.1111/j.1439-0310.1989.tb00495.x.

36 Aleksey Vnukov, "Crowboarding: Russian Roof-Surfin' Bird Caught on Tape," *YouTube*, January 12, 2012, https://www.youtube.com/ watch?v=3dWw9GLcOeA.

37 *Wikipedia*, s.v. "Marriage," https://en.wikipedia.org/wiki/Marriage.

38 Anne Jeschke, "Zu welchen Gefühlen Tiere wirklich fähig sind" (What are animals really capable of feeling?), *Welt*, February 15, 2015, https:// www.welt.de/wissenschaft/umwelt/article137478255/Zu-welchen-Gefuehlen-Tiere-wirklich-faehig-sind.html.

39 Herbert Cerutti, "Clevere Jagdgefährten" (Clever hunting associates), *NZZ Folio*, July 2003, http://folio.nzz.ch/2003/juli/clevere-jagdgefahrten.

40 Sindya N. Bhanoo, "Ravens Can Recognize Old Friends, and Foes, Too," *New York Times*, April 23, 2012, http://www.nytimes.com/2012/04/24/ science/ravens-can-recognize-old-friends-and-foes-too.html.

41 Jasmin Kirchner, Gerhard Manteuffel, and Lars Schrader, "Individual Calling to the Feeding Station Can Reduce Agonistic Interactions and Lesions in Group Housed Sows," *Journal of Animal Science* 90, no. 13 (2015): 5013–20, https://www.animalsciencepublications.org/ publications/jas/articles/90/13/5013.

42 Jeremy Hance, "Birds Are More Like 'Feathered Apes' Than 'Bird Brains,'" *The Guardian*, November 5, 2016, https://www.theguardian.com/ environment/radical-conservation/2016/nov/05/birds-intelligence-tools-crows-parrots-conservation-ethics-chickens.

43 Donald M. Broom, Hilana Sena, and Kiera L. Moynihan, "Pigs Learn What a Mirror Image Represents and Use It to Obtain Information," *Animal Behaviour* 78, no. 5 (2009): 1037–41. doi: 10.1016/j.anbehav. 2009.07.027.

44 McGill Newsroom, "Squirrels Show Softer Side by Adopting Orphans," news release, June 1, 2010, https://www.mcgill.ca/newsroom/channels/news/squirrels-show-softer-side-adopting-orphans-163790.

45 Discovery of Sound in the Sea, "How Do Marine Fish Communicate Using Sound?" http://www.dosits.org/animals/useofsound/fishcommunicate/.

46 Mathias Kneppler, "Auswirkung des Forst- und Alpwegebaus im Gebirge auf das dort lebende Schalenwild und seine Bejagbarkeit" (Effects of forest and mountain roads in the mountains on the ungulates that live there and hunting them), Abschlussarbeit des Universitätslehrgangs Jagdwirt/-in an der Universität für Bodenkultur Wien, Lehrgang VI 2013/2014, Seite 7.

47 Sebastian Herrmann, "Peinlich" (Embarrassing), *Süddeutsche Zeitung,* May 30, 2008, http://www.sueddeutsche.de/wissen/schamgefuehle-peinlich-1.830530; Christine R. Harris, "Embarrassment: A Form of Social Pain," *American Scientist* 94, no. 6 (2006): 524–33, http://charris.ucsd.edu/articles/Harris_AS2006.pdf.

48 Mary Bates, "Rats Regret Making the Wrong Decision," *Wired,* June 8, 2014, https://www.wired.com/2014/06/rats-regret-making-the-wrong-decision/.

49 CBS DFW, "Scold Them All You Want, Dogs Feel No Shame Says Behaviorist," February 26, 2014, http://dfw.cbslocal.com/2014/02/26/scold-them-all-you-want-dogs-feel-no-shame-says-behaviorist/.

50 Elsevier, "What Really Prompts the Dog's 'Guilty Look,'" *ScienceDaily,* June 14, 2009, https://www.sciencedaily.com/releases/2009/06/090611065839.htm.

51 Friederike Range, et al., "The Absence of Reward Induces Inequity Aversion in Dogs," *PNAS* 106, no. 1 (2009): 340–45. Reported by Nell

Greenfieldboyce in "Dogs Understand Fairness, Get Jealous, Study Finds," NPR, December 9, 2008, http://www.npr.org/templates/story/ story.php?storyId=97944783.

52 Jorg J. M. Massen, Caroline Ritter, and Thomas Bugnyar, "Tolerance and Reward Equity Predict Cooperation in Ravens (*Corvus corax*)," *Scientific Reports* 5, article no. 15021 (2015). doi: 10.1038/srep15021.

53 Ishani Ganguli, "Mice Show Evidence of Empathy," *The Scientist,* June 30, 2006, http://www.the-scientist.com/?articles.view/articleNo/ 24101/title/Mice-show-evidence-of-empathy/.

54 Loren J. Martin, et al., "Reducing Social Stress Elicits Emotional Contagion of Pain in Mouse and Human Strangers," *Current Biology* 25, no. 3 (2015): 326–32. doi: 10.1016/j.cub.2014.11.028.

55 Felicity Muth, "Can Pigs Empathize?" *Scientific American,* January 13, 2015, https://blogs.scientificamerican.com/not-bad-science/ can-pigs-empathize/.

56 Aleksander Medveš, "Crow Rescue," *YouTube,* June 21, 2014, https:// www.youtube.com/watch?v=gJ_3BN0m7S8.

57 Mark Matousek, "The Meeting Eyes of Love: How Empathy Is Born in Us," April 8, 2011, https://www.psychologytoday.com/blog/ethical-wisdom/ 201104/the-meeting-eyes-love-how-empathy-is-born-in-us.

58 Henry H. Lee, et al., "Bacterial Charity Work Leads to Population-Wide Resistance," *Nature* 467 (2010): 82–85. doi: 10.1038/nature09354.

59 Gerald G. Carter and Gerald S. Wilkinson, "Food Sharing in Vampire Bats: Reciprocal Help Predicts Donations More than Relatedness or Harassment," *Proceedings of the Royal Society B* 280, no. 1753 (2013). doi: 10.1098/rspb.2012.257.

60 Birk Grüling, "Ein Hund im Wolfspelz ist Tierquälerei" (A dog in wolf's clothing is animal abuse), *Zeit Online,* July 3, 2014, http://www.zeit.de/ wissen/umwelt/2014-06/tierhaltung-wolf-hybrid-hund; Patricia McConnell, "The Tragedy of Wolf Dogs," *The Other End of the Leash* (blog), July 13, 2013, http://www.patriciamcconnell.com/ theotherendoftheleash/the-tragedy-of-wolf-dogs.

61 University of Massachusetts Amherst, "Lord's Study May Explain Why Wolves Are Forever Wild, but Dogs Can Be Tamed," news release, January 17, 2013, https://www.umass.edu/newsoffice/article/lords-study-may-explain-why-wolves-are-forever-wild-dogs-can-be-tamed.

62 "Rehbock greift zwei Frauen beim Walking an" (Buck attacks two women out walking), *Schwarzwälder Bote,* April 21, 2015, http://www.schwarzwaelder-bote.de/inhalt.st-georgen-rehbock-greift-zwei-frauen-beim-walking-an.aee11194-f2cd-40b8-ba43-5204586dfc0c.html.

63 Dana Krempels, "The Mystery of Rabbit Poop," Miami College of Arts & Sciences, http://www.bio.miami.edu/hare/poop.html.

64 Jennifer Welsh, "Why Pooh Bear Loves Honey, but Tigger Doesn't," *Live Science,* March 12, 2012, http://www.livescience.com/18994-carnivores-taste-sweets.html.

65 Katsuhisa Ozaki, et al., "A Gustatory Receptor Involved in Host Plant Recognition for Oviposition of a Swallowtail Butterfly," *Nature Communications* 2, article no. 542 (2011). doi: 10.1038/ncomms1548.

66 Patricia McConnell, "Why Do Dogs Roll in Disgusting Stuff?" *The Other End of the Leash* (blog), June 1, 2015, http://www.patriciamcconnell.com/theotherendoftheleash/why-do-dogs-roll-in-disgusting-stuff.

67 Lesley A. Smith, Piran C. L. White, and Mike R. Hutchings, "Effect of the Nutritional Environment and Reproductive Investment on Herbivore-Parasite Interactions in Grazing Environments," *Behavioral Ecology* 17, no. 4 (2006): 591–96. doi: 10.1093/beheco/ark004.

68 Jenny Stanton and Jane Flanagan, "Beware, Lions Crossing (and Mauling)!" *MailOnline,* July 13, 2015, http://www.dailymail.co.uk/news/article-3159262/Beware-lions-crossing-mauling-Breathtaking-moment-beasts-catch-antelope-inches-stunned-tourists-safari-park-road.html.

69 Michael Petrak, "Rotwild als erlebbares Wildtier—Folgerungen aus dem Pilotprojekt Monschau-Elsenborn für den Nationalpark Eifel" (Red deer as watchable wildlife—results of a pilot project in Monschau-Elsenborn for Eifel National Park), in *Von der Jagd zur Wildbestandsregulierung,*

NUA-Heft Nr. 15, Seite 19, Natur-und Weltschutz-Akademie des Landes Nordrhein-Westfalen (NUA), May 2004.

70 Association for Psychological Science, "The Genetics of Fear," news release, March 9, 2009, http://www.psychologicalscience.org/news/releases/the-genetics-of-fear-study-suggests-specific-genetic-variations-contribute-to-anxiety-disorders.html.

71 Dietmar Spengler, "Genes Learn from Stress," Research Report 2010, Max Planck Institute of Psychiatry, https://www.mpg.de/431776/forschungsSchwerpunkt?c=148053.

72 Amy Liptrot, "How Berlin's Urban Goshawks Helped Me Learn to Love the City," *The Guardian,* May 13, 2005, https://www.theguardian.com/cities/2015/may/13/berlin-goshawks-urban-wildlife-tempelhof-airport-birdwatching.

73 Kathryn Westcott, "What Is Stockholm Syndrome?" *BBC News Magazine,* August 22, 2013, http://www.bbc.com/news/magazine-22447726.

74 Dietrich von Holst, "Populationsbiologische Untersuchungen beim Wildkaninchen" (Research into the population dynamics of wild rabbits), in *LÖBF-Mitteilungen* (Transactions of the Institute of Ecology, Land Use and Development, and Forestry, North Rhine-Westphalia), Nr. 1 (2004): 17–21.

75 Canadian Honey Council, "How to Make a Pound of Honey," http://www.honeycouncil.ca/chc_poundofhoney.php.

76 Nick Jans, *The Grizzly Maze: Timothy Treadwell's Fatal Obsession with Alaskan Bears* (New York: Dutton, 2005).

77 Eva Bellemain, et al., "The Dilemma of Female Mate Selection in the Brown Bear, a Species with Sexually Selected Infanticide," *Proceedings of the Royal Society B* 273, no. 1584 (2006): 283–91. doi: 10.1098/rspb.2005.3331.

78 MIT News, "Rats Dream about Their Tasks during Slow Wave Sleep," May 18, 2002, http://news.mit.edu/2002/dreams.

79 Michel Jouvet, "States of Sleep," *Scientific American,* February 1, 1967, https://www.scientificamerican.com/article/the-states-of-sleep/.

80 Hubertus Breuer, "Die Welt aus der Sicht einer Fliege" (A fly's eye view of the world), *Süddeutsche Zeitung,* May 19, 2010, http://www.suedeutsche.de/panorama/forschung-die-welt-aus-sicht-einer-fliege-1.908384.

81 Elke Maier, "A Four-Legged Early-Warning System," *MaxPlanck-Research 2* (2014): 58–63, https://www.mpg.de/8252362/W004_Environment-Climate_058-063.pdf.

82 Gabriele Berberich and Ulrich Schreiber, "GeoBioScience: Red Wood Ants as Bioindicators for Active Tectonic Fault Systems in the West Eifel (Germany)," *Animals (Basel)* 3, no. 2 (2013): 475–98. doi: 10.3390/ani3020475.

83 Katherine M. Williams, et al., "Increasing Prevalence of Myopia in Europe and the Impact of Education," *Ophthalmology* 122, no. 7 (2015): 1489–97. doi: 10.1016/j.ophtha.2015.03.018.

84 Emily Benson, "'Landmark Study' Solves Mystery behind Classic Evolution Story," *Science,* June 1, 2016, http://www.sciencemag.org/news/2016/06/landmark-study-solves-mystery-behind-classic-evolution-story.

85 Andreas Lebert and Claudia Wüstenhagen, "In Gedanken bei den Vögel" (In the minds of birds), *Zeit Online,* July 20, 2015, http://www.zeit.de/zeit-wissen/2015/04/hirnforschung-tauben-onur-guentuerkuen; Juan D. Delius, Robert J. Perchard, and Jacky Emmerton, "Polarized Light Discrimination by Pigeons and an Electroretinographic Correlate," *Journal of Comparative and Physiological Psychology* 90, no. 6 (1976): 560–71, https://www.ncbi.nlm.nih.gov/pubmed/956468.

86 Paul E. Miller and Christopher J. Murphy, "Vision in Dogs," JAVMA 207, no. 12 (1995): 1623–34, http://redwood.berkeley.edu/bruno/animal-eyes/dog-vision-miller-murphy.pdf.

87 Sarah Bordenstein, "What Is a Tardigrade?" *Microbial Life Educational Resources,* http://serc.carleton.edu/microbelife/topics/tardigrade/index.html.

88 *National Geographic,* "The Human Condition: Our Imprint Deepens as Consumption Accelerates," http://www.nationalgeographic.com/earthpulse/human-impact.html.

89 Gesamtverband der Deutschen Versicherungswirtschaft, "Zahl der Wildunfälle sinkt auf 247.000" (The number of accidents involving wildlife declines to 247,000), July 10, 2014, http://www.gdv.de/2014/10/zahl-der-wildunfaelle-sinkt-leicht/.

90 Peter Werner and Rudoph Zahner, "Biological Diversity and Cities: A Review and Bibliography," Bundesamt für Naturschutz, *BfN-Skripten* 245 (2009).

91 Michael Slezak, "Ancient DNA Suggests Dogs Split from Wolves 40,000 Years Ago," *New Scientist,* May 27, 2015, https://www.newscientist.com/article/mg22630235-500-ancient-dna-suggests-dogs-split-from-wolves-40000-years-ago/.

92 Stefan Müller, "Interkulturelles Marketing" (Intercultural marketing), PowerPoint presentation, 2007, slide 4, "Paraverbale Kommunikation" (Paraverbal communication), https://tu-dresden.de/gsw/wirtschaft/marketing/ressourcen/dateien/lehre/lehre_pdfs/Mueller_IM_G1_Kommunikation.pdf?lang=de; Jeff Thompson, "Is Nonverbal Communication a Numbers Game?" *Psychology Today,* September 30, 2011, https://www.psychologytoday.com/blog/beyond-words/201109/is-nonverbal-communication-numbers-game.

93 Max Planck Institute, "'Look at That!' Ravens Use Gestures, Too," news release, November 29, 2011, https://www.mpg.de/4664902/ravens_use_gestures.

94 Elodie F. Briefer, et al., "Segregation of Information about Emotional Arousal and Valence in Horse Whinnies," *Scientific Reports* 4, no. 9989 (2015). doi: 10.1038/srep09989.

95 Inken De Wit, "Wie Pferde Emotionen äussern" (How horses express emotion), *ETH Zürich,* May 15, 2015, https://www.ethz.ch/de/news-und-veranstaltungen/eth-news/news/2015/05/wiehern-nicht-gleich-wiehern.html.

96 The Gorilla Foundation, http://www.koko.org.

97 Danny D. Steinberg, Hiroshi Nagata, and David P. Aline, *Psycholinguistics: Language, Mind, and Word* (London: Longman, 1982), 150; Roc Morin, "A Conversation with Koko the Gorilla," *The Atlantic,* August 28, 2015, http://www.theatlantic.com/technology/archive/2015/08/koko-the-talking-gorilla-sign-language-francine-patterson/402307/.

98 Jane C. Hu, "What Do Talking Apes Really Tell Us?" *Slate,* August 20, 2014, http://www.slate.com/articles/health_and_science/science/2014/08/koko_kanzi_and_ape_language_research_criticism_of_working_conditions_and.html.

99 Dictionary.com, s.v. "Soul," http://www.dictionary.com/browse/soul.

100 Thomas Goschke, "Kognitionspsychologie: Denken, Problemlösen, Sprache" (Cognitive psychology: thinking, problem solving, speech), Module A1: Kognitive Prozesse (Cognitive processes),PowerPoint presentation, 2013, https://tu-dresden.de/mn/psychologie/allgpsy/ressourcen/dateien/lehre/lehreveranstaltungen/goschke_lehre/ss2013/folder-2013-04-15-9955666685/vl01_einfuehrung?lang=en.

101 Pol Marquer, Teresa Rabade, and Roberta Forti, "Pig Farming Sector—Statistical Portrait 2014," Eurostat *Statistics Explained,* http://ec.europa.eu/eurostat/statistics-explained/index.php/Pig_farming_sector_-_statistical_portrait_2014#Production_of_porkmeat.

INDEX

abandonment, fear of, 61–62

Adams, Richard: *Watership Down*, 188

adaptation, 68–69. *See also* evolution

adoption, 11, 15–17, 110, 169, 187

aggressiveness, morality of, 191–97; bear attacks on humans, 193; bees, 191–93; male bear attacks on cubs, 195–96; morality of aggressiveness, 194–95, 196–97; rabbits, 196–97; wars and fights between animals, 194

aging, 211–16; death, 215–16; dementia, 213–15; dogs, 213–14; fear of in horses, 211, 212–13; female deer, 213

agriculture: animals killed by tractors, 143, 182; electric fences, 170; pasturing animals in woodlands, 58; pig feeding, 107; resistance to animal emotions from, 245

albatrosses, 136–37

altruism, 130–33; bacteria, 130; birds, 131; definitions and meaningfulness, 130, 131; free will and, 130, 132–33; as selfish, 132; vampire bats, 131–33

amygdala, 28–29

animals: introduction and approach to, 1–3, 30, 248; aggressiveness and morality, 191–97; aging, 211–16; altruism, 130–33; animal intelligence and willingness to eat, 41–42; in artificial environments, 224–29; body temperature, 72–80; brain structure, 27–36;

DAVID SUZUKI INSTITUTE

THE DAVID SUZUKI INSTITUTE is a non-profit organization founded in 2010 to stimulate debate and action on environmental issues. The Institute and the David Suzuki Foundation both work to advance awareness of environmental issues important to all Canadians.

We invite you to support the activities of the Institute. For more information please contact us at:

David Suzuki Institute
219 – 2211 West 4th Avenue
Vancouver, BC, Canada V6K 4S2
info@davidsuzukiinstitute.org
604-742-2899
www.davidsuzukiinstitute.org

Cheques can be made payable to The David Suzuki Institute.